MONKEY FARM

Robert M. Yerkes at his desk. Courtesy of the Robert Mearns Yerkes Papers. Manuscripts and Archives, Yale University Library.

MONKEY FARM

A History of the Yerkes Laboratories of Primate Biology, Orange Park, Florida, 1930–1965

Donald A. Dewsbury

Lewisburg
Bucknell University Press

Associated University Presses
2010 Eastpark Boulevard
Cranbury, NJ 08512

The paper used in this publication meets the requirements of the American National Standard for Permanence of Paper for Printed Library Materials Z39.48-1984.

Library of Congress Cataloging-in-Publication Data

Dewsbury, Donald A., 1939–
Monkey farm : a history of the Yerkes Laboratories of Primate Biology, Orange Park, Florida, 1930–1965 / Donald A. Dewsbury.
p. cm.
Includes bibliographical references and index.
ISBN 0-8387-5593-3 (alk. paper)
1. Primates—Research—Florida—Orange Park—History. 2. Yerkes Laboratories of Primate Biology—History. I. Title.

QL737.P9D49 2006
599.8′072′075919—dc22

2005014999

PRINTED IN THE UNITED STATES OF AMERICA

Contents

Illustrations

Tables

Preface

THE MAIN TITLE OF THIS BOOK, "MONKEY FARM," REFERS TO THE popular vernacular name used by Orange Park residents in referring to the station. The facility was founded as the Laboratories of Comparative Psychobiology of Yale University by Robert M. Yerkes, one of the leading psychologists of the twentieth century. Yerkes' goal was to secure knowledge that could be applied in a program of social engineering for the betterment of humanity. The Yerkes Laboratories became the largest and most important collection of chimpanzees for research in the world. During its thirty-five-year history it was home to some of the leading behavioral scientists of the time.

Writing its history has been a delight and a challenge. My graduate education was in experimental psychology and I had a twenty-five-year career as a comparative psychologist interested in the evolution of social and reproductive behavior. When I decided to switch from laboratory research to historical research I began with a series of relatively small projects, as I cut my teeth in a new field. Eventually, I sought a subject that was amenable to a more substantial, book-length treatment. The Yerkes Laboratories of Primate Biology seemed to be an ideal topic. The facility was located about a one-and-a-half-hour drive from the University of Florida. It was the home to some of the leading scientists in the field and became a leading research facility for work on nonhuman primates. I felt that most topics were within my range of expertise. Further, as I investigated the history of the Laboratories, it became clear that the story was both significant and rich enough to merit book-length treatment. There are many facets to the laboratory, the research that was done there, and the people who worked there. I became fascinated with its story.

I began work on the project during the early 1990s but I did not seek a book contract at that time for several reasons. I felt that I did not yet have sufficient credibility as a historian of psychology to merit strong support. Further, this was a new project for me and I was uncertain as to how the project would unfold and what paths it would take. Finally, I take deadlines seriously and I did not want the pressure of a fixed deadline. This turned out to be a good decision. Because of other responsibilities, I was able to work on the writing of the book only a couple of months each summer. Although I continued to gather material and conducted related work during the academic year, my other commitments and projects prohibited intensive work on the book. Finally, in the fall semester of 2002, I took the first sabbatical of my career to complete a version of the manuscript with which I felt comfortable.

After contacting a number of publishers, I was fortunate to receive a contract from Bucknell University Press and Associated University Presses for this book. However, it was necessary to reduce the length of the original manuscript substantially. During the summer of 2004, I cut approximately half of the text of the 2002 manuscript. I had to eliminate a lot of pertinent information about the Yerkes Laboratories but believe that in the process I have produced a much more readable book. I am placing a copy of the complete (more comprehensive but less readable) 2002 manuscript in the Archives of the History of American Psychology at the University of Akron, so that the more detailed story will be available to scholars wanting to probe more deeply.

I have used a variety of sources in preparing this manuscript. In all, I have thus created from approximately fourteen cartons of material a coherent story. The material comes primarily from three sources that merit mention. First, much of the material about the actual research that was conducted comes from the published record in scientific journals and books. The Yerkes Regional Primate Research Center maintains a nearly complete list of its publications that was useful in tracing the published record. Other useful information has come from libraries and from newspaper and magazine accounts.

Various archives and special collections served as a second resource. I wish to thank the many archivists and other personnel at the Yale University Library; the Archives of the American Philosophical Society in Philadelphia, the Archives of the History of American Psychology in Akron, Ohio; the Woodruff Library of Emory University; the Yerkes Regional Primate Research Center; the University of Florida Special Collections Department; the Pennsylvania State University Library; the Archives of the National Academy of Sciences; and the Rockefeller Archive Center. I am grateful for a travel grant for work at the Rockefeller Archive Center in New York state. Many of the materials that are now in the Woodruff Library were obtained by me at the present Yerkes Laboratories with the free access and photocopying provided through the great courtesy of then-director Frederick A. King and Ms. Cathy Yarbrough. Their successors, Thomas Insel and Stuart Zola, have also been of help. I was able to trace Karl Lashley's stepdaughter (Paul Schiller's daughter), Christina Schlusemeyer, to a horse farm in Ocala, Florida and obtained from her many useful documents, including Karl Lashley's Vanuxem Lectures later published by Orbach (1998). She has been most helpful and has since arranged for their donation to the University of Florida Special Collections Department, for which we are most grateful.

The third kind of information has been through interviews. I interviewed most of the living scientists who worked in Orange Park. I also interviewed several of the support staff from the Orange Park years. In addition, I interviewed numerous residents of the Orange Park area for background concerning the context. I thank all of those who gave their time as well as numerous other resources. At the risk of omitting some names, I list them here. Among my interviewees were: Gershon Berkson, Irwin S. Bernstein, Roberta Yerkes Blanshard, Meredith P. Crawford, Donald Eitzman, John Falk, Louis Gainers, Betty Green, Paul M. King, Mary Jo Cotney McTammany, William A. Mason, Emil W. Menzel, Jr., Vincent Nowlis, Lelon J. Peacock, Robert Pollard, Karl H. Pribram, Austin and Helen Riesen, Arthur J. Riopelle, Jimmy and Judy Roberts, Charles and Ginga Rogers, Christina Schlusemeyer, Samuel A. Spencer, Johannes W. and Margarete Tigges, Marvin Wilhite, Rene Wilkie, Jane Elder Wulff, and David Yerkes. Several scientists, including William A. Draper, Theodore A. Jackson, Ronald J. Schusterman, and S. D. S. Shirley Spragg provided audiotapes describing their experiences. Most of the interviews and tapes have been transcribed. Others, including Marcia Huey Anderson, Alta M. Belyea, Janeth Carpenter, Justina Draper, W. Richard Dukelow, Marilyn Ferster Gilbert, Norman B. Guilloud, George Harrell, Naosuke Itoigawa, Dora Nissen Mann, Jan Moor-Jankowski, Cathy Hayes Nissen, Jack Orbach, Sue V. Saxon, Wallace I. Welker, and Jane Elder Wulff, provided written or photographic material that was most helpful. Mrs. Roberta Yerkes Blanshard was especially gracious in hosting my stay at her retirement home in Hamden, CT. The Menzels were similarly hospitable during my visit to their Long Island home. A highlight of my efforts was sailing on Charlie and Ginga Rogers's fine sailboat among the tall ships off the Maine coast. Among those with whom I conversed only by telephone were Murray E. Jarvik, Broun

Parmenter, Richard T. Smith, and Helen B. Warren. I thank the College of Liberal Arts and Sciences of the University of Florida for the sabbatical leave during the Fall 2002 semester that enabled me to complete the earlier version of this manuscript.

I thank Douglas K. Candland for his help in facilitating acceptance of the manuscript at Bucknell University Press and Dr. Greg Clingham for working with me to develop a plan for publication. I must especially thank Dr. Ludy T. Benjamin, Jr. for reading the complete, longer manuscript and making suggestions for reduction that were both insightful and helpful. His careful reading and constructive criticism played an important role in making my manuscript more readable. Gail Erny read the entire revised manuscript and page proofs and provided many helpful suggestions. Laura Mestayer Rogers prepared the index. I thank the people at Associated University Presses for their work on this manuscript.

The process of writing a book is time-consuming, stressful sometimes and invigorating others. Those around me, Bryan, Laura, Gail, and Joyce, have been tolerant and supportive throughout. Theirs has been yet another important contribution to this book and to my life.

Donald A. Dewsbury
Gainesville, Florida

MONKEY FARM

1
Introduction

On the surface, June 1930 was a terrible time to begin a new enterprise. It had been less than a year since the American stock market had collapsed and a worldwide financial crisis was raging. Jobs were lost and cutbacks were everywhere. Nevertheless, that was the time that Yale University's Robert Mearns Yerkes, undaunted by all that surrounded him, was finally able to bring to fruition a dream and a plan on which he had worked for many years: a remote facility for the study of chimpanzees. Local residents, not well versed in the taxonomy of primates, would call it the "monkey farm."

Around the year 1900 what was to become the Yerkes Laboratories of Primate Biology (YLPB) in Orange Park, Florida, was only a vaguely formed idea in Yerkes' mind as he was studying psychology at Harvard University. The path to fruition was a long and difficult one. However, the Orange Park facility would become the leading center for research on the great apes in the world and would survive in Florida until 1965, when the development of a federal program of primate research centers led to a reorganization of the facility and a move to the campus of Emory University in Atlanta, Georgia. This is the story of the Orange Park facility from 1930 to 1965.

The Life of an Institution

One might wonder whether an institution can have a life and thus be the subject of a biography. In at least some senses it can. Like a human life, the Yerkes Laboratories in Florida had a long gestation, birth, growth, maturity, and, depending on one's perspective, a death or reincarnation. Like a human life, the story of its history is a story of a balance between stability and change.

David Osterbrock (1997), independent of my effort, used similar life-span language in writing a history of a different institution bearing the

Yerkes name: the Yerkes Observatory. There is no relationship between either the two Yerkes families or the two institutions that I have been able to discover.

An institutional history can be written from various perspectives: one would be that it was simply a scientific institution formed to gain more knowledge of our primate cousins, with its original goal being to better understand ourselves and improve society. The goals changed yet the institution grew. From another perspective, the institution is only a set of buildings. One can visit the buildings today and see only remnants of what was built there. Now one sees a health food store and a real estate office where there was once a caretaker's cottage and a laboratory. The buildings have been transformed, but, in some sense they remain unchanged, at least for now. There were the chimpanzees—thinking and feeling animals brought there against their will in the interest of scientific investigation. In another sense, the story of the laboratory is a story of the hundreds of people who worked there and whose lives were often shaped there. Their commitment to science was important. Beyond that, however, there are stories of births and deaths, joy and tragedy. The stories sometimes take on the air of a soap opera, with divorce, suicide, intrigue, and tragic loss by drowning or skiing. All are important to a balanced understanding of the facility and my endeavor here is to view the YLPB from the various perspectives in order to gain that understanding.

Themes to Be Explored

Although this is the biography of an institution, and primarily a narrative history, it also serves as a vehicle to explore several themes and thus illustrate some important processes in the history of science in the twentieth century.

Individuals

A long-standing debate in the history of science and related fields concerns the extent to which events are the result of the actions of "great persons," traditionally "great men," or of broader cultural determinants. Historical currents of varying breadth had great impact on events related to the YLPB. On the other hand, within these broad trends, the influences of individuals were substantial.

THE DIRECTORS

The five directors set the tone for the whole facility. There was the stern Robert Yerkes, full of plans and convinced that the path to human understanding and progress lay in understanding chimpanzee behavior. There was the uncompromising and austere Karl Lashley, working to understand brain function and how brains could be changed to alter behavior. Henry Nissen was the extremely competent man but he was quiet, self-effacing, and not born to be an administrator. Arthur Riopelle, a man for the new scientific environment, followed Nissen and was poised to transform the laboratories from their somewhat sleepy nature to meet the new realities of visibility in a changed funding environment. The ever-colorful Geoffrey Bourne was the entrepreneur who would oversee the final transformation and the move from Florida to the urban center of Atlanta.

STAFF SCIENTISTS AND SUPPORT STAFF

While the directors were the generals, each with his own critical role in shaping the changing directions of the laboratories, it was the scientific staff members who were responsible for realizing the plans of the directors. They came to Orange Park with their own backgrounds and goals, which interacted with those of the director and the facility, sometimes positively and sometimes negatively. Most would build careers elsewhere on foundations built in Orange Park.

At the bottom of the hierarchy were the support staff—the caretakers, bookkeepers, secretaries, and technicians. The facility could not function without them. Yet their stories are rarely told. Directors and scientists write autobiographies and sit for interviews; secretaries and caretakers rarely do. Thus, it is difficult to locate much detail on their lives. Nevertheless, they were important.

THE ANIMALS

The role of the nonhuman primates was at least as important as that of any human who worked there. In the interests of healthy development, they were separated from their mothers at birth, confined, and endured occasional surgery, irradiation, and lots of bad weather. They performed in complex learning experiments, demonstrated their social proclivities, and, above all, reproduced. They delighted in their own antics and mischief. In many ways, human and chimpanzee lives were intertwined. There was joy both at the birth of a baby and at the birth of Alpha, the first chimpanzee to be born there, and there was the sadness at the loss of both human and chimpanzee lives.

CHANGE

Surely, the period from 1930 to 1965 was a time of great change in many respects. How various scientists handled change was an important determinant of their success. Some found change threatening; others embraced it. Those who failed to change with the times put their careers at risk.

The Scientific Milieu

PATRONAGE

A principal theme in this book concerns changing patterns of patronage and the impact of these changes on the research that would be conducted at a facility such as the YLPB. At the turn of the century there were few sources of significant support for scientific endeavors; "big science" was impossible. When private foundations established by industrial moguls, began to appear during the early part of the century, it provided a funding base sufficient for the emergence of a different kind of science. Money from private sources was available to support a limited number of major research facilities and projects.

By the middle of the century the federal government entered into the business of supporting science to such a degree that the foundations that had caused much of the growth paled by comparison. In each of these phases there were pressures that affected the sort of individuals who would be given support and the research that would be supported. Investigators altered their proposals to conform with the ideals that would garner funding. One could argue that the greatest changes in funding patterns in the history of science occurred within the time span covered by the YLPB, thus the facility was a microcosm of how changing funding patterns affected science throughout the nation.

PROFESSIONALIZATION

A related theme throughout the text is that of the increasing professionalization of both science in particular and academia in general during this period. There is the popular image of the scientist as the lone investigator, the individual iconoclast at work in his (gender reference deliberate) solitary laboratory. Yet, because there are many different kinds of scientists, generalizations are risky. Nevertheless, the prefoundation scientist could be more easily characterized as an individualist of the sort described above than could the scientist of later years.

With the availability of foundation funding it became important that the aspiring scientist was also an entrepreneur. One needed to meet the foundation officers on their terms and possess the social skills to convince them of the merits of one's projects. Not all took kindly to the new order. Psychologist James McKeen Cattell caricatured the scientific beggars at the doors of the rich: "Scientific men still wait hat in hand on the almoners of Mr. Carnegie and Mr. Rockefeller. University professors even yet find silence to be silver, flattery to be gold" (Cattell 1922, 572).

The role of the scientist-administrator would become even more complex with the shift from foundation to federal funding. Administrative details often became so onerous that many leading scientists would remain chained to their desks, rarely visiting their own laboratories, which often were the purview of layers of assistants. As the demands of a laboratory director's job changed, the kind of person who could best fulfill that job changed.

What started out as an often individual endeavor became a highly social one with administrative and business skills essential. The foundations came to value cooperation, by which they meant more than casual collaboration among neighboring scientists.

THE HEGEMONY OF MEDICAL RESEARCH

Society has always favored research that could be applied in ways immediately apparent to both the public and the politician. This has been the case especially since federal funding of research had increased. The effects of these societal pressures will be apparent in the story of the YLPB, particularly at its end.

ETHICS OF ANIMAL RESEARCH

When, in recent years, one contemplates research on nonhuman primates, the issue of ethics arises immediately. Major activism and high-profile protests have generated much attention and concern. The reader will be struck by the lack of such activity during the life of the Yerkes Laboratories in Orange Park. Such concerns existed before 1930 but had little effect during the YLPB's lifetime.

The Broader Milieu

ORANGE PARK

Many of the scientific staff experienced culture shock in Orange Park. During the early years, high-powered, highly intelligent Northerners accustomed to life in major universities and urban centers experienced life in the rural South. Orange Park was hot, full of insects, and removed from much that was taken for granted elsewhere. Some staff thrived in the laid-back atmosphere; others suffered in perceived isolation.

URBANIZATION

Another influence on the Yerkes Laboratories came from urbanization. Primate laboratories have generally been built in remote areas where land is inexpensive, there are few humans to bother the animals, and there are few humans to be bothered by the noise of the animals. During the life of the YLPB the area around it grew tremendously in conjunction with the overall urbanization of the United States. What had been a remote location in 1930 was a crowded one in 1965. The YLPB would be moved to Atlanta, a major city, where urbanization was even greater. With the problem already apparent in 1965, Emory University established a field station of the Primate Center in Lawrenceville, Georgia, then a remote area. (That too has been developed and now is a community of commuters.) From 1930 to 1965, it became increasingly difficult to keep the laboratory separate from the surrounding population, alas, with consequences for both.

RACE AND GENDER

One could not write about a southern institution during this period without considering race. Throughout its history in Florida, there was a clear differentiation of roles with white scientists supervising black workers. Early on, the arrangement was generally accepted and issues of race are rarely made explicit. These issues became more prominent during the 1960s when change was in the air and increasing discomfort with the status quo became expressed.

Occupants of the upper echelons of the laboratory administration were all male. Indeed, the first director, Robert Yerkes, would employ no female scientists. That changed soon after Lashley succeeded him. Some important women worked in the laboratories after that. Nevertheless, they were in the minority as they were in much of science at the time.

BROAD PHILOSOPHIES

Broad philosophical outlooks were important to the history of the YLPB. For example, the progressivism of both Robert Yerkes and the Rockefeller Foundation were essential to the founding of the laboratories. It was a time when the prospects for human improvement seemed bright, and there was great optimism that science would provide the route to such improvement.

May (1959) portrayed Americans around the turn of the twentieth century as sharing three primary articles of faith in a national credo: a practical idealism, a belief in progress, and a respect for culture. It was an era when Americans believed, with little doubt, in a set of enduring values of civilization. The course of both evolution and history was viewed as inevitably upward, and it was thought that we might be able to provide a boost to that process. The center of culture, thus conceived at the time by some, was in the Boston area and its chief arbiters were Robert Yerkes' teachers: Josiah Royce, William James, George Santayana, and others at Harvard. These ideals infused much of the work of the foundations and much of the thinking of Harvard's Robert Yerkes.

This milieu changed during the period after World War I. The values that had been thought enduring came to be regarded as ephemeral and open to question (Hawley 1992; May 1959). Yerkes was a product of the earlier time, out of sync with the new mentality. Yet he used these values to structure his views of chimpanzees and their sexual and family relationships.

WORLD WAR II

Surely, the most important historical event that occurred during the life of the YLPB was World War II. It would affect the laboratories, as it affected the whole country. Housing and the loss of personnel to the war effort were important effects of the war.

Historiography

In recent decades there has been a dramatic shift within the discipline of the history of science. As elsewhere in academia, there was an increased professionalization. The field of the history of science was formerly dominated by older scientists who, past their scientific primes, wrote history. Often the histories that they wrote glorified the past through a selective retelling of the accomplishments of the "great men," "great dates," and great ideas, shorn of their failures and complexities.

Increasingly, however, a cadre of scholars of varying ages developed the "new history," which reached the history of psychology much later than it did other areas of history (e.g., Furumoto 1989). The new historians have become specialists in the field of the history of science. They have examined historical events more objectively and have striven to write critical, rather than ceremonial histories. By turning to original sources, especially unpublished materials in archival and manuscript collections, they have obtained more balanced understandings of historical events and have placed historical events in broad social and cultural contexts.

Some of the new historians have gone too far. In reading their work one sometimes gets the view that scientists are but corks bobbing on the inevitable waves of cultural change. Some attacks on the objectivity of science portray science as a tool of the power structure bent on containing the lower classes in a Marxist struggle. Scientists are portrayed as Machiavellian creatures obsessed with struggles for power. The very concept of objective truth, central to many scientific approaches, has suffered many telling attacks.

Because histories, like scientific research programs, are constructed by individuals (rather than discovered in mature form) the perspective of the author is essential for understanding a work. My personal history is thus relevant. In brief, I had an active career as a comparative psychologist and so have a strong sympathy with the field. On the other hand, in recent years my primary effort has been in the history of psychology. I have benefited from the perspectives of the various students of science studies including the new historians, cultural studies advocates, students of the sociology of scientific knowledge, and postmodernists.

My approach is thus a blend of my scientific and historical interests and training with a sincere effort at balance. Like my comparative psychology, my history is, above all else, eclectic. I take what I can from wherever I can. I will be internalist and externalist, celebratory and critical, past-minded and presentist where I think it will help to flesh out the story. I want to understand the YLPB and to tell its story. To do that I must present events which include both successes and failures, as I understand them. Humans act heroically on some occasions and stupidly on others. My goal is to be critical but fair. My sympathies are on the side of science, but only a science that is understood as culturally embedded and not above criticism. If I succeed, I shall probably offend both hard-nosed scientists and postmodernists equally. I cannot, however, pretend that I am without my own biases, as much as I strive to minimize the effects of those that are transparent to me.

The Organization of the Book

The book is organized into sixteen chapters. Most fall naturally into one of the five directorships: those of Robert Yerkes, Karl Lashley, Henry Nissen, Arthur Riopelle, and Geoffrey Bourne. Within each section I discuss the various aspects of the life of the laboratory during that administration. This includes such topics as funding, administration, personnel, the physical plant, the research that was done, and life in Orange Park at that time. A chronology of the history of the laboratories is presented in Table 1.1.

Chapter 2 provides biographical information about the founder, Robert M. Yerkes. In chapter 3, I follow Yerkes' tortuous path as he sought support for a primate laboratory, and I provide contextual information regarding early primatology, sources of funding, and Orange Park itself, to help situate the YLPB. Chapters 4 through 7 deal with the activities at the YLPB during Yerkes' tenure as director. These include, among others, difficulties of funding, personnel, and research as Yerkes tried to impose his hierarchical management style on the laboratories. Chapters 8 through 10 cover the time of Lashley's directorship. This was a period of greater individual freedom for investigators and for some staff members who would go on to international renown. The range of species was expanded and a reductionistic science was emphasized. In chapters 11 and 12, I cover the period when Henry Nissen and interim director Lelon Peacock sat in the director's seat. This was a challenging period as many adjustments were required under changing conditions. Among the key events occurring during this period were the transfer of ownership to Emory University and the arrival of a controversial program of Skinnerian operant conditioning. The next two chapters, 13 and 14 deal with activities during the directorship of Arthur Riopelle as he tried to develop the facility in Orange Park to take advantage of new funding and research trends. During this period the Emory officials worked, over Riopelle's strenuous objections, to use new federal funds to transfer the laboratories to their home campus in Atlanta. In chapter 15, I describe the YLPB during the directorship of Bourne, as preparations were made to move the laboratories. This was a period of little innovation and few new personnel. In chapter 16, I conclude by briefly discussing events that transpired in Orange Park and Atlanta after the facility was moved. I return to many of the themes first stated in chapter 1 in an effort to provide closure.

Table 1.1.
Chronology of the Yerkes Laboratories of Primate Biology

1876	Robert Yerkes born
1877	Town of Orange Park founded
1902	Yerkes Harvard PhD
1905	Robert Yerkes and Ada Watterson marry
1916	Yerkes *Science* paper proposes a primate station
1917	World War I; Yerkes to Washington
1923	Chim and Panzee to New Hampshire
1924	Yerkes moves to Yale as Professor of Psychology
1924	Yerkes to Cuba with Madame Abreu
1925	Yerkes' *Almost Human*
1925	Rockefeller Foundation appropriates $10K annually for four years for Yale primate laboratory
1925	Yale primate facility opened
1925–28	Yerkes studies Congo (gorilla) in Florida
1929	Yerkes and Yerkes' *The Great Apes*
1929	Rockefeller Foundation funding for $25K feasibility study
1929	Rockefeller Foundation funding $500K
1930	Station opened as the Laboratories of Comparative Psychobiology, Yale University
1935	Anthropoid Experiment Station of Yale University becomes the Yale Laboratories of Primate Biology
1935	Yale Laboratories of Primate Biology incorporated in Florida
1937	Rosalia Abreu Memorial Fund started by Pierre S. Abreu
1938	Yerkes asks to be relieved of Directorship
1938	Yale establishes committee to study the past and future of the laboratories
1939	Rockefeller grants $189K for 1939–44 support
1939	Digging begins for new building
1940	Henry Nissen initiates Infant Studies Program
1941	Yerkes retires as Director
1942	Lashley named new Director
1942	Laboratories renamed in honor of Robert Yerkes
1942	Northern branch discontinued
1943	Yerkes' *Chimpanzees: A Laboratory Colony*
1947	Viki home rearing study begins
1949	Paul Schiller dies
1949	Robert Blum drowns at Vilano Beach
1951	*The Ape in Our House* published
1954	Nissen-Hayes wedding
1954	Nissen and Pribram program
1954	Viki dies

continued

Table 1.1. *Continued*

1954	Operation on Lashley
1955	Lashley resigns as Director; Nissen becomes Director
1956	Robert Yerkes dies
1956	Emory takes over laboratory operations
1957	Lashley–Claire Schiller wedding
1957	Nissen spinal injury
1957	NSF Committee visits YLPB
1958	Lelon Peacock Acting Director
1958	Nissen suicide
1958	Lashley dies
1959	Arthur Riopelle new Director
1962	Rumors confirmed that YLPB may move to Atlanta
1962	Emory gets funds for Regional Primate Center
1962	Riopelle announces his move to Delta Center
1962	Riopelle finishes as Director
1963	Geoffrey Bourne becomes Director
1963	Ada Watterson Yerkes dies
1965	Animals moved to Atlanta
1976	Marvin Wilhite announces plans to develop site

2
Robert M. Yerkes, Founder

THE YERKES LABORATORIES OF PRIMATE BIOLOGY WERE FOUNDED as the result of the efforts of one man, Robert Mearns Yerkes, working as part of an academic-social-foundation network. The laboratories bore the clear stamp of their founder and a consideration of his life is necessary to understand some characteristics of the facility. I shall present a sketch of some aspects of his life treading lightly on those parts that belong to the history of the YLPB because those events will be covered in later chapters (see also Dewsbury 1996, 2000). Robert Yerkes was, in many respects, a very private and formal man, lacking in the extroverted character of many successful administrators. Yet, he was a hard worker who planned endlessly and was able to negotiate the political climate with success. Few would call him brilliant; rather, he was consistent, persistent, and, in some matters, insistent.

EARLY LIFE

Robert Mearns Yerkes was born on a farm in Breadysville, Bucks County, Pennsylvania on May 26, 1876, "the centennial year of our republic" (Yerkes 1950, 1). He was a first-born; the oldest child of Silas Marshall Yerkes and Susanna Addis Carrell Yerkes. There were two sisters, one died at age three, and two brothers, who were so much younger that he described his childhood as lonely.

Yerkes recalled happy childhood days with "out-of-door amusements; free, unrestricted, unaided study and enjoyment of nature; the care of household and farm pets; the capture and taming of wild animals" (Yerkes 1930a, 381). Many biologists and comparative psychologists had such early contacts with animals.

Yerkes' mother was "a woman of rare sweetness of disposition and unusual ability, beloved of all who knew her" (382). Although he recalled pleasant interactions with his grandparents, he did not get along well with his father: "except for my attachment to my mother and the influences of a capable, level-headed young German then in the employ of my father, I probably should have run away from home before my fourteenth year" (382). An early bout with scarlet fever had lasting effects on his endurance. For his part, Yerkes noted that he had "been described as a moody, strong-willed, unsuggestible child, difficult to control" (383). Observing a cousin who was a physician, Robert Yerkes daydreamed of becoming a physician too, an occupation quite incompatible with his father's wish for him to remain in agriculture. Yet being around his cousin had a beneficial effect on his behavior and ambition.

Yerkes began school in his eighth year, delayed because of shyness and scarlet fever. He described himself lacking in musical ability, artistic ability, and proficiency with languages. His teachers tried to change the left-handedness of their young student. Characteristically, he resisted all of their efforts.

At the age of fifteen, Yerkes was sent to the State Normal School at West Chester. The next year he entered Ursinus Academy. Although his father lacked the means to provide him with such education, an uncle, a homeopathic physician, allowed him to earn his way by doing chores around his property. The next year he was admitted to Ursinus College where his studies emphasized chemistry and biology. Like many students at that age, his college experiences led him to question the religious beliefs of his parents: "the study of biological science and of philosophy laid the foundation for a revolt against religious dogmatism and the seeming partialness and inadequacy of the creed to which I had been born" (Yerkes 1950, 78). He graduated from Ursinus in 1897.

Harvard

Although he had planned to attend medical school in Philadelphia, a growing wish to get broader educational experiences brought Robert Yerkes to Harvard with a provisional undergraduate classification in 1897; the Ursinus degree was not fully accredited. In his first year, he took courses primarily in English, philosophy, and animal biology. He received an A. B. degree and graduate status after that year.

Yerkes decided to continue in graduate school rather than medical school. In biology he had the opportunity to work with a number of influential professors, including George H. Parker, William E. Castle, and C. B. Davenport. The latter is especially interesting. Kevles (1985) credits

Davenport with introducing Yerkes to the eugenics positions of Sir Francis Galton; Yerkes later espoused similar views. However, it is not clear if that affected his primate work.

As Yerkes was interested in both psychology and biology, his friend and advisor, Josiah Royce, suggested comparative psychology, as it combined the two. Hugo Münsterberg agreed that there were opportunities in animal psychology and the switch was made in 1899.

Comparative psychology as a discipline was new, having descended from such scientists as Charles Darwin, C. Lloyd Morgan, and George John Romanes in England. William James's views, strongly influenced by evolutionary theory, established the framework within which Yerkes could work at Harvard. When Robert Yerkes was a student at Harvard, comparative psychology was still a developing field but one that was already recognized as a valid part of psychology as a whole. Yerkes would play a major role in its development.

In psychology at Harvard, Yerkes was influenced especially by James, Royce, and Münsterberg, but also by George H. Palmer and George Santayana. During his first three years, however, he spent much time in the Zoological Laboratory. Yerkes received his doctorate in 1902 with a dissertation on the sensory reactions and physiology of the nervous system in jellyfish.

Yerkes came out of his shell in Cambridge, with its plays, operas, and concerts. They were a revelation to the farm boy; the intellectual environment was vibrant. He continued to grapple with matters of religion (Yerkes 1950). He spent several summers at the Woods Hole Biological Station on the Massachusetts coast. In that environment he gradually became more comfortable around women. While there, Yerkes conducted a number of studies on fiddler crabs, jellyfish, and medusae, and studied habit formation in turtles and crabs. Yerkes' early studies set the tone for later work at the YLPB. Although he worked with invertebrates and so-called lower vertebrates, he studied sensory function, reactions to simple stimuli, and habit formation (e.g., Yerkes 1899, 1901) (see Hilgard 1965).

On Münsterberg's recommendation, Yerkes stayed on as an instructor in comparative psychology after graduation. He described his years on the Harvard faculty during 1902–17 as "living on cream" (Yerkes 1950, 108). He had ample time for research and continued to work with jellyfish during summers at Woods Hole.

Yerkes' first book, *The Dancing Mouse: A Study in Animal Behavior* (1907), became a classic. It is a summary of a comprehensive and well-planned analysis of the behavior of mutant house mice, with a description of the history of the population, development, naturally occurring behavior, sensory processes, learning and habit formation, and the inheritance of behavior. This list of topics is one that Yerkes would often use and is close to his major framework for work at the YLPB a quarter of a century later.

The Harvard years continued with a growing publication record, including both empirical studies and articles of greater breadth, that earned increasing respect from Yerkes' colleagues around the country. An extensive collaboration with John B. Watson led to a classic paper on the methods for detailed study of vision in animals (Yerkes and Watson 1911). He also published a textbook in psychology (1911), bringing him a strong national and international reputation. He was elected president of the American Psychological Association (APA) for the year 1916–17.

During this time Yerkes, with the help of Gilbert Hamilton, devised a new testing apparatus that he believed might reveal ideational behavior in animals more fully than did the devices used previously. With the "multiple-choice method," he studied a variety of species and found, for example, that pigs could solve some problems that baffled crows and rats (e.g., Yerkes 1916d).

It was during his Harvard years, while "living on cream," that Robert Yerkes married and started his family. One summer at Woods Hole, he met Ada Watterson, a botanist from Cleveland who was teaching at Columbia University. They married on September 16, 1905, and agreed facetiously that when Ada produced a baby, Robert would produce a book. Roberta Watterson Yerkes arrived along with *The Dancing Mouse* in 1907, and David Norton Yerkes along with *Introduction to Psychology* in 1911. Among other achievements, Roberta would become an editor for the Yale University Press and marry the philosopher Brand Blanshard. David would become a successful architect.

Like many of his colleagues, Yerkes liked to escape from Cambridge during the summers. Woods Hole provided the ideal location for some years. However, in 1911, as both their family and the town of Woods Hole grew, the Yerkes family purchased a farm in Franklin, New Hampshire, near that of their friends, Dr. and Mrs. Walter B. Cannon. The Pleasant Hill Farm became the "Franklin Field Station" and would serve as a refuge and a research station for many years (Yerkes 1914c).

In the meantime, Yerkes' interest in primates was growing. As will be discussed later, he began active campaigning for an anthropoid station in 1913. Yerkes learned of Wolfgang Köhler and the station at Tenerife,

Canary Islands, and planned to go there on a half-year's leave with his family around 1912. But the start of World War I upset those plans, so the Yerkes family spent February–August, 1915, on sabbatical leave with Gilbert Hamilton in Montecito, California, where Yerkes conducted a number of studies on the primates maintained at Hamilton's Montecito station. He worked with an irus macaque, a rhesus macaque, and an orangutan named Julius, and produced his first substantive publication on primate behavior. Yerkes was surprised that the monkeys did worse on the multiple-choice tasks than did the pigs. He thought he had produced "the first curve of learning for an anthropoid ape" (Yerkes 1916e, 131). Yerkes' many observations led him to conclude that he had evidence of ideation, roughly the equivalent of Köhler's "insight," in the orangutan. This work and Köhler's at Tenerife occurred at about the same time, although Köhler's, started a bit earlier, was more prolonged than Yerkes'. Communication between the two scientists was difficult because of the German-American hostilities and, as a result, there were some hard feelings between them. Köhler felt his work had been slighted. These differences were eventually patched up (Ash 1995; Haslerud 1979).

Despite his accomplishments, things were not going well between Robert Yerkes and the Harvard administration. Yerkes' problems there were based primarily on questions concerning the place of animal research in the philosophy department at Harvard and were a significant distraction for him throughout his time in Cambridge (Dewsbury 1992; O'Donnell 1985; Wight and Smith 1998). It was only with Münsterberg's effort that he was initially appointed as an Instructor, rather than Teaching Fellow, and he retained that title for six years. Since President Lowell was not sympathetic to comparative psychology, and Münsterberg's support was inconsistent at best, Yerkes was finally promoted only with the understanding that he would emphasize educational psychology in his teaching more than he had done previously. Characteristically, Yerkes resisted the efforts to move him from comparative psychology, the area in which he felt most comfortable. However, he did begin work in psychopathology and testing with his friend E. E. Southard at the Boston Psychopathic Hospital. The work continued until Yerkes left Harvard in 1917. At Boston Psychopathic, Yerkes and his colleagues developed a "Point Scale," an intelligence test that provided an alternative to the Binet scale, which had recently become standard.

In 1915 Yerkes declined an offer to move to the University of Minnesota and reorganize the psychology department there. But when the offer was reactivated the next year, and it was apparent that things at Harvard would not improve, Yerkes accepted the Minnesota offer.

World War I and Washington

World War I

In 1917, when President Woodrow Wilson announced American entry into World War I, Yerkes assumed leadership, as APA president, in an effort to bring psychology to the service of the war effort. He chaired an APA Committee on the Psychological Examination of Recruits. This would lead him into the most controversial phase of his life and one that has received much attention from historians of psychology. He became an officer in the Sanitary Corps of the U.S. Army and spearheaded the development of a testing program for recruits. Yerkes and his associates developed the Army Alpha test for literate recruits and the Army Beta test for those unable to read. Historians generally agree that the work had a more favorable effect in establishing psychology as a viable science than on the conduct of the war, but the inference that this was by design seems unsupported (Dewsbury 1996).

Most controversial were the conclusions drawn by Yerkes and his associates after the war. They compiled a long report (Yerkes 1921a) that Yerkes (1923a) summarized and interpreted. Hereditarian and former student of C. B. Davenport, Yerkes concluded that: "If we may safely judge by the army measurements of intelligence, races are quite as significantly different as individuals. . . . Almost as great as the intellectual difference between negro and white in the army are the differences between white racial groups" (Yerkes 1923a, 363–64). There is no escaping the racist implications of his conclusion. Yet Yerkes and his associates were products of their time. By the mid-twenties, Yerkes had reexamined the data and appeared to back off some of his views (Samelson 1979).

Robert Yerkes personified the progressivist ethic of his time. Although today we view some of their positions as retrogressive, the progressives of Yerkes' era thought of the testing effort as providing opportunity based on ability rather than social class or privilege.

Washington

After the war Yerkes elected never to report for academic duty in Minneapolis. He had been active in the work of the National Research Council since 1917 and chose to resign his Minnesota professorship to stay on in Washington as chairman of the NRC's Division of Research Information. He would remain in Washington from 1919 until 1924, serving in a number of capacities, including as chairman of the Committee on Problems of Human Migration and of the Committee for Research in Problems of Sex (CRPS) (Clarke 1998; Pickren 1997). Although not all of these efforts were fully successful, Yerkes gained much administrative experience. Another beneficial result for his later career was the opportunity to make important contacts with influential administrators such as George Ellery Hale, John C. Merriam, Frank R. Lillie, and James Rowland Angell.

Back to Primates

Robert Yerkes was biding his time until he could bring his dream and plan for an anthropoid station to fruition. In the meantime, he took advantage of several opportunities to gain experience working with primates.

While on vacation in Franklin, New Hampshire in the summer of 1923, Yerkes received a letter from the director of the New York Zoological Park, William T. Hornaday, informing him of the availability of two chimpanzees that were temporarily housed at the zoo. The owner would only sell them as a pair. The female was a so-called "common" chimpanzee (*Pan troglodytes)* in bad health and of little value. The male, however, was recognized as a rare specimen, what we now know as a bonobo (*Pan paniscus*). The asking price was $3,000 for the pair at a time when the typical price for a chimpanzee was $200–$500 (Yerkes 1963). Yerkes offered $2,000 for the pair, a bold move for a man relying on personal funds and with a lifetime savings of under $3,000. The owner, Noel Lewis, agreed and brought them to Franklin in August; the bonobo was named Chim and the chimpanzee, Panzee. Both did well in Franklin, but in September, they had to be moved to Washington and Panzee survived only a few months. Much of the story of Chim and Panzee can be found in the book *Chimpanzee Intelligence and its Vocal Expressions* by Yerkes and Blanche W. Learned (1925). Of particular interest in this book is the method of representing chimpanzee vocalizations in standard musical notation in the age before sound spectrograms. Yerkes supplemented the work at Franklin with observations made on the three adult chimpanzees at the Bronx Zoo in New York City.

In the summer of 1924 Yerkes went to Havana to work with the primate collection maintained by Madame Rosalia Abreu. Yerkes had first learned of Madam Abreu's collection in 1915 with the birth there of the first chimpanzee born in captivity. Yerkes was accompanied by his associates Harold C. Bingham and Josephine Ball, as well as "Prince

Chim." The story of the research done that summer at Quinta Palatino is told in Yerkes' book *Almost Human* (1925a).

A respiratory epidemic occurred in the Abreu collection that summer and Chim died. Although Madam Abreu insisted on giving Yerkes $500 as compensation, he noted that "it could not buy another Chim. I never have seen another like him" (1963, 208). Robert Yerkes' father died while he was in Cuba, but Robert elected to stay in Havana rather than return for the funeral.

Yerkes soon acquired four more chimpanzees; two were purchased for $400 each from animal collector John L. Buck in 1925. With a touch of irony, as the famous Scopes "monkey trial" was being conducted in Tennessee during July of that year, Yerkes named the male Bill, after William Jennings Bryan, prosecutor in the case. The female was named Darwina (or "Dwina"), in honor of Charles Darwin. Later that summer Yerkes bought a pair of young chimpanzees (Pan and Wendy, named for Peter Pan characters), from a ship's officer in New York for $500. Both would live long lives in the Yale Laboratories.

The four chimpanzees got along well, although the initial two pairs of chimpanzees tended to persist as pairs for some time. One is struck with the similarity between that summer of 1925 as lived by the chimpanzee quartet, and Yerkes' own youth in Bucks County: "our New England pasture was unique, with chimpanzees swinging from its birches or watching with timid curiosity the herd of grazing cows" (1950, 270). Photographs of the foursome at their dinner table are among the most engaging, at least in an anthropomorphic sense, of the many that Yerkes published. (See page 56.)

Yerkes was put into contact with Ben Burbridge of Jacksonville, Florida, owner of a young mountain gorilla, Congo, and was subsequently invited to Jacksonville to study him. Eventually, he was able to study Congo for six to eight weeks during each of three successive winters—a nice break from winter in New Haven! Congo was kept at the home of Ben's brother, James, near South Jacksonville, not far from Orange Park. After Burbridge sold Congo to John Ringling for his circus, Yerkes was able to study Congo during 1927–28 at the Ringling winter headquarters in Sarasota, Florida. The results of this work were reported in three monographs on *The Mind of a Gorilla*, one covering each year (1927b, 1927d; 1928a). In the first year, Yerkes concentrated on a set of problems, such as stick problems, delayed response problems, box stacking, and hidden stick problems. During the second and third winters his emphasis of study was developmental changes as Congo matured. Yerkes became convinced of the animal's ability to display insight.

Of special interest are the conclusions Yerkes drew from his observations and the language he used to describe them. In an era in which the behaviorism of his friend, John B. Watson, was becoming highly influential, Yerkes' sounded more like that of a personality theorist. Yerkes wondered about "biological advantages and disadvantages of extroversion and introversion" (1927b, 152). The orangutan displays "less evidence of optimism and impulsiveness" (180). Over successive years, Congo became easier to test: "what was first described as aloofness, self-sufficiency, and introversion, came finally to be recognized as the directive and inhibitory expression of timidity, lack of self-assurance, and mistrust" (1928a, 76). Throughout the reports, Yerkes described Congo's behavior in terms of such human traits and mental states. But Yerkes was out of sync with the prevailing behaviorism studies' climate in attributing anthropomorphic characteristics to his primate subjects. This may have been one cause of his later troubles.

Another important milestone in Robert Yerkes' pre-Orange Park career as a primatologist was the production of the massive review of extant knowledge of apes, *The Great Apes: A Study of Anthropoid Life* (1929), written by Robert and Ada Yerkes and dedicated to their children. The book begins with a treatment of ancient knowledge of the apes and an historical treatment of increasing knowledge of them. There are then sections on gibbons, orangutans, chimpanzees, and gorillas, each including information on the anatomy, life history, social behavior, intelligence, and other characteristics of the different apes. Characteristic of a Yerkes publication, there are many photographs of animals, and of buildings; there are also some illustrations drawn by David Yerkes. The completion of this comprehensive and authoritative summary may have been a factor in convincing the Rockefeller Foundation of the scientific validity of Yerkes' plan for an anthropoid station.

To Yale

Many academics who enter such administrative positions never return to their academic pursuits. Yerkes, however, wanted all along to return to research, especially his work with nonhuman primates. His "escape" from Washington was arranged through his Washington friend James Rowland Angell, who had assumed the presidency of Yale University. Yerkes, along with Angell, developed a plan for an Institute of Psychology at Yale that included a position for Yerkes as a Professor of Psychology with no teaching duties. The plan was funded by the Laura Spelman Rockefeller Memorial and, in 1924 Yerkes declined a job offer from the University of California and was off to New Haven with an intermediate stop in Havana. The details of Yerkes' life in New Haven and Orange Park, and much that happened in the ensuing years, will be discussed later. I now skip to the post-Orange Park years.

Robert Yerkes' Later Career

Retirement

In 1941, at the age of 65, Robert Yerkes retired as director of the Orange Park Laboratories; however, he continued as a Professor of Comparative Psychobiology at Yale until he reached the mandatory retirement age of 68 in 1944. He remained active in professional affairs during retirement. He completed a final published book, *Chimpanzees: A Laboratory Colony* (1943), a semipopular version of the story of the YLPB. Yerkes' book was initially to have been cowritten with Henry Nissen, but Nissen was unable to complete the job and Yerkes completed it himself.

Although he naturally was somewhat less active in the mobilization effort during World War II than he had been during World War I, Yerkes played a pivotal role in the organization of efforts by psychologists to serve in the war effort. Psychological services were in greater demand during the second war than during the first. Yerkes served on the Emergency Committee in Psychology of the National Research Council and chaired its Subcommittee on Survey and Planning. Harvard psychologist Edwin G. Boring wrote: "Here Yerkes was at his best—always wise, always eager to explore new possibilities, never ready to surrender when no means seemed to be available or success seemed improbable" (Boring, 1956, 139). A critical problem on which the subcommittee focused was the lack of unity among psychologists. Disenchanted with the treatment given applied psychology in the older APA, the applied psychologists had formed their own organization, the American Association of Applied Psychologists (AAAP). Numerous other, smaller groups also came into existence. Lack of coordination would impair the effectiveness of psychology and psychologists in the war effort. Yerkes' subcommittee therefore proposed that an Intersociety Convention be called to bring greater unity to psychologists and their war efforts. The convention was held in May 1943 in New York with Robert Yerkes as the temporary chair. The convention rewrote the APA bylaws making a reunification of psychology possible—at least for the time being. In the view of Dael Wolfle: "Although many people helped plan the APA reorganization, and many did the work necessary to make the plans effective, one name stands out as the person who started the movement and who gave wise counsel in the months when the new constitution was taking shape. That is Robert M. Yerkes. The new constitution reflects the will of the membership; otherwise it would not now be in effect. But the membership has Professor Yerkes, more than any other one person, to thank for its new constitution" (Wolfle 1946, 3).

Yerkes saw alternatives for his retirement years. The lure of experimentation had weakened for him, and he rejected further research with animals. He did maintain contact with the activities at Orange Park and with some of the scientists there. An especially interesting correspondence, including at least sixty-three letters between 1949 and 1956, developed between Yerkes and Cathy Hayes (later Cathy Nissen) in Orange Park (Dewsbury 1994b). He rejected the alternative of writing technical volumes on the work at the YLPB. He reflected: "that some younger person, not so closely associated with the organization and conduct of our research, might do a better job than I. History writing is not my forte; nor is the reporting and interpretation of the work of other investigators" (Yerkes 1950, 353).

He chose the third alternative, "systematically studying my own life history as experience and achievement, and by fitting together the fragments" (353). The result was his autobiography, *The Scientific Way,* or more informally, *Testament* (1950), an excellent source of information about the life of Robert Yerkes. The principal theme developed is the thesis that: "there are ways of life which are potentially better than the religious. Among them is the way of knowledge and enlightenment, truth seeking and willingness to carry responsibility instead of casting it upon the infinite" (iv). The manuscript was rejected for publication by several publishers, including Yale University Press. Yerkes' daughter Roberta, an editor there regretfully participated in the rejection of her father's autobiographical manuscript. In part of the manuscript, Yerkes had returned to themes with which he had struggled earlier in his life. The emphasis on religion did not deal with the major themes of Yerkes' life and presented little that was really new (Blanshard 1993). Yerkes was "considerably shocked" (Yerkes 1950, vi) when his second major revision and condensation of the manuscript brought such criticism from his daughter. Thus the last major project of Robert Yerkes' life was never published.

Honors

Robert Yerkes received many honors. He was a member of the National Academy of Sciences of the United States and of the American Philosophical Society. He was a fellow of the American Academy of Arts and Sciences. Yerkes was APA president from 1916–17 and president of the American Society of Naturalists in 1938. He received honorary doctorates from Ursinus College and Wesleyan University. A bust of Yerkes was placed in the Darwinian Museum in Moscow. In January 1954 he was awarded the Gold Medal of the New York Zoological Society (Hilgard 1965).

Robert Yerkes: The Man

Traits and Abilities

Robert Yerkes spent an important part of his research career in the study of individual differences among humans and seemed equally interested in the taxonomy of individual differences in some of the primate species. He was especially interested in his own abilities and characteristics in relation to those of others. Some consideration of them is critical to understanding the history of the YLPB.

He believed himself to be of but average intelligence: "probably if my intelligence quotient had been determined at ten years it would have been not less than 125, but again I feel confident that in most respects my intellectual abilities approximated the average" (Yerkes 1950, 25). In this he may have been underestimating himself. Yerkes could not have accomplished all that he did with average abilities. Clearly, however, he did not possess the razor-sharp intellectual quickness that characterized some of his colleagues.

Yerkes noted that his memory was not exceptional and he was not good at accumulating facts. He turned this into a plus, however: "Is it perhaps true that persons of exceptional retentive memories tend to become encyclopedically learned, whereas those of relatively poor memories, among whom I undoubtedly should number myself, tend rather to become inventive, inquiring, and constructive?" (1930a, 385).

Above all, Yerkes was a planner. He wrote: "I do not remember a time when my day's activities were not self-planned and within permissible limits also self-determined" (1950, 25). Yerkes was not always an easy man with whom to work. He related this to his love of planning: "whether it be a merit or a shortcoming, I am not a good follower. It cramps my dominant trait, planfulness, and reduces me to a species of intellectual slavery" (1930a, 406).

Yerkes was not socially gifted. He repeatedly refers to his bashfulness, especially when he was younger. Although his adult experiences brought him out of his shell to some degree, as late as 1950 he would recall that "later in life my behavior often was mistaken for a pose or affectation" (1950, 29). With regard to his work as a teacher, he noted that "I lacked the inclusive and highly retentive memory, the ready wit and quick reaction, the sympathetic interest in others and easy personal adjustment which are so important for the enjoyment of teaching and for eminent success" (1950, 131).

He impressed others as being very serious. "A caustic scientist put the idea another way, 'I just can't believe that anybody can feel as serious as Yerkes looks'" (Williams 1947, 27). Another remarked that "Yerkes lacked all humour" (Zuckerman 1978, 68) and "that he was a much drier and more austere character than his letters had given me to suppose" (69). One scientist noted, "I've never seen him hook his elbow over a bar and tell a joke" (Williams 1947, 27). Kevles noted: "shy and stiffly formal, Yerkes was not the academic politician who could glad-hand his way to the presidency of a learned society" (1968, 566). However, even though he was not an extrovert, he loved parlor games, frequently entertained laboratory personnel and friends, valued family pleasures, and was indeed capable of laughter.

Yerkes was not a drinker. Nowlis (1994b) recalled that "for a while he took a glass of ginger ale before he went to bed. But he found that he had a tendency to erotic thoughts when he had ginger ale, so he gave up the ginger ale."

Hard work characterized much of Yerkes' career; "love of work and the power to tap new reservoirs of energy seem to have been paternal heritages which the circumstances of my life greatly strengthened" (Yerkes 1930a, 404). This tendency to work hard at all he undertook was one of the reasons he was so effective on committees: "how Yerkes did love the committees he served on, and how hard he worked at them!" (Elliott 1956, 493). "Loyalty was the central fact about him: intense devotion to his wife, daughter and son, faithfulness to his many friends—mostly fellow scientists—and passion for service in the causes in which he believed" (494).

Robert Yerkes was a gentleman, in the traditional sense of the word, and he expected others to be likewise. An example from Nowlis (1994b) illustrates this: "We were in Constitution Hall for the APA General Session where elections were announced. . . . and I was sitting with him and we were right in back of Harlow, and when it was announced that Harlow was the new president, Harry [Harlow] stood up and shouted in a very uncouth way 'it's about time.' And Yerkes just turned beet red at this breech of decent conduct" (3).

Values

Robert Yerkes sometimes imposed his own values on the observations that he made. He often was faced with the need to interpret a behavioral pattern in the absence of full information and so, of necessity, he filled it in. In this respect, scientists are no different from other mortals.

Old-fashioned principles of the golden rule and fair play were important to him (Yerkes 1950, 341). Highest among the expressed satisfactions of Robert Yerkes' life were "(1) work, (2) marital and familial companionship and affection, (3) social usefulness and social status" (139). Writing in *Testament* (1950), he ranked such values as wealth, power, fame, popularity, and personal beauty as low or negative. A case can be made, however, that his ego was stronger than suggested by these self-serving passages.

Yerkes, in contrast to more recent understanding of chimpanzee social organization, viewed the family, more specifically, the core family that was predominant in his time and place, as the core of chimpanzee society (Haraway 2000). This came from Yerkes' own strong sense of family. He wrote "emphatically I reaffirm my belief in the family as social institution" (Yerkes 1950, 138) and "I do not know of a single separation or divorce among my ancestors or near relatives, including the nineteen marriages of my uncles and aunts" (7).

Yerkes' views of women were typical for a man of his time; he believed that "women are more deeply concerned with the perpetuation of the species than are men; more wrapped up in the problems and chores, privileges and satisfactions of housekeeping" (296). Today, it seems remarkable that one could write of the "privileges and satisfactions of housekeeping." He believed that "from birth educational practices should be adapted to sex as well as to individual characteristics" (235). His attitude toward women would affect his decisions in adding staff to his facilities. When Eleanor Gibson, later a distinguished psychologist, visited Yale in an effort to work in his laboratory, she was summarily dismissed (Gibson 1980, 246).

Robert Yerkes' comments, made in 1950, on racial differences are interesting in light of his history of work in the testing movement and the controversial conclusions with which he had been associated: "I shall never cease to be grateful for my parent's [*sic*] catholicity of view about human differences. It very early taught me to judge persons by their actions instead of their station in life, education, racial origin, or color" (1950, 18).

On Science

Yerkes' views on science were a blend of his values of duty and disinterested science. Science could be used to improve the human condition and to find a better way of living: "to say that I was intent on the perfecting of man and his way of life seems an improvement on the statement that I was devoted to the promotion of research" (242). Yerkes was a believer in "*mental engineering,* as exemplified in the applications of psychological science to the practical problems of education, vocational guidance, and the industrial and fine arts" (168–69). "By 1925 experience had convinced me that man must understand and learn to control himself, as well as his living and inanimate environment, in order to live adaptively and to further or use, instead of working against, evolutionary processes" (274). He sought "the emergence of a community or brotherhood of man under a world federation" (225) and saw science in the service of this ideal. Haraway (1989) stressed the role of control in Yerkes' program; with proper knowledge, nature and society could be managed. She also detected strong themes of paternalism, domination, and power in his writings. It is not clear that Yerkes' goals were effectively realized in Orange Park.

Yerkes' science was integrative rather than reductionistic. He struggled with the issue of naturalistic versus experimental methods in science and tried to blend the two. An experimentalist who believed in the importance of testing animals under controlled conditions, he knew that this must be done with a full understanding of the animal's natural proclivities; "every experimentalist in animal psychology should be also a naturalist whose love of animals and sympathy with them enables him to understand their behavior" (Yerkes 1909, 4). "First of all, the observer of an animal under experimental conditions should be thoroughly familiar with the habits and instincts, the sensitiveness and fears of his subject" (5).

Late in his career he summarized his values and beliefs in his oft-cited creed. It is a statement that might be regarded as odd in the twenty-first century but was reflective of a man of his time:

I believe:

In knowledge of the natural order as basis of man's life.

In the supernatural—soul, spirit, absolute—as possible.

In religious experience as awareness of superindividual influence or being.

In man's responsibility for his life but not for eternity, destiny, immortality.

In the obligation of man to strive to guarantee to every individual the inalienable right to be well born and well reared.

In the dignity and perfectibility of man as part of the natural order.

In the worship of ideal manhood rather than godhood and of manliness rather than saintliness.

In usefulness through fellow service.

In the natural origin of conscience, morality, and codes of human conduct.

In the priority of life over death, effort over prayer, knowledge over faith, and reason over wishfulness. (Yerkes 1950, 425)

3
The Road to Orange Park

ROBERT YERKES (1943, 289) RECALLED THAT "SOMEWHERE AROUND 1900, while I was in the Graduate School of Arts and Sciences, Harvard University, the notion of a special research institute for comparative psychobiology came to me." Over time, this goal, fused with his belief in the importance of work with monkeys and apes, became an ideal, that of a facility for the study of primate psychobiology. Yerkes later spelled out his plan in a pivotal *Science* magazine article: "There should be provided in a suitable locality a station or research institute which should offer adequate facilities (1) for the maintenance of various types of primates in normal and healthy condition; (2) for the successful breeding and rearing of the animals to many generations; (3) for systematic and continuous observation under reasonably natural conditions; (4) for experimental investigations from every significant biological point of view; (5) for profitable cooperation with existing biological institutes or departments of research throughout this country and the world" (1916c, 231). Much of Yerkes' career can be seen as motivated by this long-term goal of establishing such a primate research facility.

EARLY PRIMATOLOGY

Yerkes' efforts can be understood in the context of the primatology that existed in his time. Histories of early knowledge of the great apes can be found in Yerkes and Yerkes (1929) and Mitchell (1999). Among the early students of primates in zoos were Britishers Darwin, Romanes, and L. T. Hobhouse.

Primates in Early Comparative Psychology

The first studies of nonhuman primates in American comparative psychology occurred soon after the turn of the century when a cluster of studies appeared. A variety of issues concerning intelligence, reasoning, problem solving, and imitation were contested. Watson (1908) studied imitation in a baboon, a capuchin, and two rhesus monkeys but found little evidence of true imitation. Andrew Kinnaman of Clark University, conducted a series of tests (1902) on two rhesus monkeys, concluding that he had found evidence of higher orders of imitation and of general notions and reasoning. Lightner Witmer, who founded the first psychological clinic in the United States, studied several species of primates, including some from vaudeville shows. He was especially interested in problems of language. Witmer (1909, 205) was prophetic in writing: "I venture to predict that within a few years a chimpanzee will be taken early in life and subjected for purposes of scientific investigation to a course of procedure more closely resembling that which is accorded the human child." That was exactly what would be done in Orange Park.

Robert Yerkes' entry into primate research appears to have been through the efforts of his student, Melvin E. Haggerty, who began working with three cebus monkeys at Harvard in 1907 and completed his doctoral dissertation on imitation at the New York Zoological Park (Haggerty, 1909). W. T. Shepherd and Shepherd Ivory Franz also studied primates during this period.

Surely the influence of Darwinism, the central place accorded by Romanes' writings on mental continuity, and the "new psychology," with its emphasis on controlled tests and measurement, were important in the upsurge of interest in the study of primates. Comparative psychology was developing and the study of nonhuman primates was near its center.

Early Primate Research Stations

Probably the first such facility was the Primate Station, Tenerife, Canary Islands, founded in 1913. The idea of founding a station for the observation of anthropoid apes originated with Max Rothmann, a Berlin physician (Ash 1995; Teuber 1994). The most famous research done at Tenerife was that of Gestalt psychologist Wolfgang Köhler and included tasks such as box stacking, pole use, detour experiments, and the use of sticks to secure incentives (Köhler 1925). Köhler came to believe that the animals solved the problems with sudden flashes of insight, rather than by a gradual process of trial-and-error learning. The station was formally closed in October 1920.

The Pasteur Institute of Kindia was founded in 1923 at Kindia, French Guinea by the Pasteur Institute of Paris for the utilization of various species of monkeys and apes in medical research (Lapin and Fridman

1966; Yerkes 1929a). The staff had been unsuccessful in trying to breed chimpanzees in captivity. Yerkes developed a plan to send an investigator there to conduct a systematic study of chimpanzee life under natural conditions and then to bring back to the United States some young chimpanzees for further study (Yerkes 1931a). This was accomplished by Henry Nissen (1931) and formed the core of the original colony in Orange Park.

The Sukhumi Breeding Station (now the Institute of Medical Primatology of the Russian Academy of Medical Sciences) was opened in 1927 (Lapin and Fridman 1966). Although many animals at the station died, the managers succeeded in slowly increasing the population and celebrated the birth of a hamadryas baboon in 1928. The story of the gradual development of methods useful in keeping primates makes one realize how difficult it was and the extent of Robert Yerkes' achievements in this regard.

Private Collections

In addition to these government-supported facilities, a number of private collections were important in the progress of scientific research. Madam Rosalia Abreu developed a significant private collection at her family home, Quinta Palatino, in Havana, Cuba. On April 27, 1915 she was able to report the first birth of a chimpanzee, Anuma, in captivity—an important step for the idea of a station for the study of the great apes. By the time that Robert Yerkes and his entourage arrived to study her animals in the summer of 1924, the colony numbered nearly eighty primates. After Madam Abreu died in 1930 (Bourne 1971; Hahn 1988), some of her chimpanzees were donated to help populate the Orange Park collection.

Gilbert VanTassel Hamilton, a physician who had done graduate work at Harvard, served at the McLean Hospital in Worcester, Massachusetts, where one of his patients was financier Stanley McCormick. When McCormick was released from McLean he engaged Hamilton as his personal physician on his estate at Montecito, near Santa Barbara, California. Hamilton soon built a small comparative psychology laboratory and developed a small primate colony on the McCormick estate (Hamilton 1911). In 1914, the colony included three types of old-world monkeys and an orangutan. Hamilton's relationship with the McCormicks ended abruptly in 1917, when it was concluded that his treatment of McCormick had been inappropriate.

A Russian, Sergei Voronov, conducted research in endocrinology at the physiological station in Paris, trying to transplant gonads from monkeys to humans. In 1922 Voronov began construction of a private monkey laboratory near Menton on the Franco-Italian border; the laboratory was completed in 1926. He maintained a colony of about 100 monkeys and chimpanzees. The laboratory of the "doctor of rejuvenation" apparently was closed as the result of military activities in 1940 (Lapin and Fridman 1966).

Another significant line of research was the program of Nadezhda Ladygin-Kohts in Russia. In a work dedicated to Robert Yerkes, Kohts (1928) described a series of investigations of rhesus macaques in puzzle boxes. In other research in the early 1920s, she studied the development of emotions, and perceptual and cognitive abilities in a chimpanzee, Ioni, and even compared his development with that of her son (Ladygina-Kohts 1935/2002).

The Origin of Yerkes' Plan

It is not clear where Yerkes' c. 1900 idea of a research station for psychobiology came from. Clearly, it evolved over time. Recall that Yerkes spent his summers at the Marine Biological Laboratory at Woods Hole, one of the premier biological stations in the world both then and now. Perhaps he extrapolated from the reality of a successful biological station to the conceptualization of a similar facility for psychobiology.

This was a time when speculative thinking about big research facilities was "in the air." When Andrew Carnegie founded the Carnegie Institution of Washington (CIW), scientists were invited to submit proposals (Kohler 1991). *Science* editor, psychologist James McKeen Cattell (1902), touched off the discussion with his invitation for suggestions for facilities. Among those proposing facilities were psychologists Hugo Münsterberg (1902) and Edward Titchener (1902).

At around the same time, Yerkes' friend John B. Watson (1906) proposed the establishment of an experimental station for the study of comparative psychology. Watson's station would house a variety of species and be focused especially on the development of behavior. Watson (1907) described for Yerkes the remarkable spontaneity of behavior of the monkeys he brought to Bird Key in the Dry Tortugas, Florida, and contrasted it with the behavior they displayed in his laboratory. This may have influenced Yerkes.

Yerkes' initial plan seems to have been for a facility for research in comparative psychobiology broadly defined (Yerkes 1915b; 1950). By the time he started actual lobbying in 1913, after Haggerty's work, the plan had become focused on primates. Clearly, then, there were many forces at work.

Patronage of Science in the Early Twentieth Century

Large, private foundations supporting scientific research emerged during the early part of the twentieth century. The YLPB, in turn, were, at first, supported by these foundations and were dependent on them for their very existence. Thus, it is important to understand both the development of these foundations and their subsequent decline in relative importance as federal sources of funding developed after World War II.

The support of science by the private foundations developed from the progressivist ideal of both understanding and curing medical and social dysfunction in society. The utility of science as the means of providing rational methods for intervention and social control was demonstrated during World War I, and "during the 1920s the amplified symbiosis between academe and the foundation resulted in a 100–fold jump in research grants to universities" (Kay 1993, 29).

With minimal outside support and modest support from within universities, nineteenth-century science had been conducted along modest dimensions. Foundation support in the twentieth century altered the ways in which science was conducted and eventually led to the development of "big science." This also meant a reduction in the independence of the individual scientist, who would now have to satisfy a benefactor. This was a development that was strongly resisted by many scientists.

The story of the various agencies created by men such as Andrew Carnegie and John D. Rockefeller, Sr. is complex, as they tended to create multiple units, sometimes with overlapping and poorly defined responsibilities. The would-be scientific entrepreneur had the advantage of several sources to which to turn, but was faced with the disadvantage of a frustrating array of units that may or may not be interested in any one project. Robert Yerkes faced exactly these problems when he set out to establish his station for research on anthropoid apes.

Andrew Carnegie's Foundations

Steel tycoon Andrew Carnegie founded the Carnegie Institution of Washington (CIW), designed to support "the exceptional man," in 1902 (Kevles 1977; Kohler 1991). Carnegie created a number of other philanthropic entities as well. The Carnegie Corporation of New York, founded in 1911, had a broader mission. Carnegie served as the first president of the corporation. James Rowland Angell served as president during 1920–1921, when he became president of Yale University (Nielsen 1972). President Angell's foundation connections would later play an important role when he served as the Yale University president at the time of the founding of the YLPB.

The Rockefeller Philanthropies

Four generations of Rockefellers created "a fleet of foundations" (Nielsen 1972). Standard Oil's John D. Rockefeller, Sr., and his associates founded several entities early in the century. The Laura Spelman Rockefeller Memorial, established in 1918, came under the directorship of Beardsley Ruml, who was trained as a psychologist and formerly associated with Angell, in 1923. Under his direction the Memorial developed a broad plan in support of the social sciences. It would provide support for Yale's Institute of Psychology.

It was the Rockefeller Foundation (RF), founded in 1909, that would eventually provide the primary support for the YLPB. The RF was dominated by a progressivist spirit in which science and medicine were viewed as the best routes to the improvement of the human condition.

The various foundations were consolidated during the late 1920s. The importance of the RF was thus greatly increased, as it took over activities related to medical science, medical education, agriculture and forestry, social sciences, the humanities, and the arts. The RF programs would be unified with a coherent goal: "it was to be essentially one program, directed to the general problem of human behavior, with the aim of control through understanding" (Fosdick 1952, 142).

The Rockefeller merger was consummated and announced on January 3, 1929 (*New York Times* 1929). The RF executive committee approved the initial allocation for Yerkes' anthropoid station on January 22, 1929. Thus, the allocation came directly from the beginning of a more general policy shift at the RF. Yerkes later reflected "had our request been delayed by even a year or two, the decision almost certainly would have been unfavorable. By this narrow margin we escaped indefinite delay or failure" (1950, 290).

Under mathematician Warren Weaver a "Science of Man" program in the Division of Natural Sciences of the RF was developed and bore the label of "psychobiology," a label also preferred by Yerkes. However, as the 1930s progressed, with no change in stated goals, Weaver moved his program in more molecular directions. The philosophy was that there was a "unity of life" and that in studying basic phenomena we develop principles that are broadly applicable. The emphasis came to be upon those parts of biology and psychology that could gain most from physics and chemistry—in short, the more physiological and reductionistic approaches. These moves would work against Robert Yerkes in the late 1930s.

Other Private Foundations

Several other foundations would be important in the support of the YLPB. These included the Ford Foundation (Nielsen 1972), the Samuel S. Fels Fund (Samuel S. Fels Fund 1963), and the Josiah Macy, Jr. Foundation (Rappleye 1955).

The Committee for Research in Problems of Sex

The Committee for Research in Problems of Sex (CRPS) of the National Research Council, an offshoot of the National Academy of Sciences, was founded in 1922 and supported with RF funds (Aberle and Corner 1953; Pickren 1997). Yerkes was elected the founding chair of the committee, a post he would hold for twenty-five years. The budget for the committee was just over $6,000 in 1921–22 and $44,000 in 1946–47, with a high point of $74,000 in 1931–32. Although these are not large sums by today's standards, the funds were critical in establishing research on a variety of problems related to sex, including some of the work of both Robert Yerkes and Alfred C. Kinsey.

The Social Sciences Research Council

The Social Sciences Research Council (SSRC), another product of Rockefeller philanthropy and the first social science institution anywhere in the world, was created in 1923 (Fisher 1993). One example of its programs was the well-known work of Winthrop N. Kellogg at the YLPB, rearing the chimpanzee Gua with his son in his home, done while Kellogg was on an SSRC fellowship.

Federal Agencies

Later, the development of federal programs had major consequences for the conduct of science in general and the YLPB in particular. Shortly after World War II, funding for science from the National Science Foundation, the National Institutes of Health, the Atomic Energy Commission, and the U.S. military would mushroom and become the primary source of funding for basic research. All would have some impact on psychology and the YLPB. (Appel 2000; Bush 1945–90; Dewsbury 2005).

Perspective

Both the Rockefeller Foundation and Robert Yerkes can be seen, in part, as products of their times. The founding of the Rockefeller and Carnegie entities "expressed Americans' faith in the efficacy of research and in a science-based social intervention" (24). Kay related this to the hegemony of capitalist ideology, a view of "American exceptionalism" (i.e., that the destiny of America differs from that of the rest of the world), and of norms of behavior derived from American Protestantism. Research was viewed as a way to get to the root causes of social dysfunction in order to enable corrective measures to be taken.

The gain of substantial and increasingly necessary support for science provided by the foundations was offset by the loss in independence of the individual scientist. For Cattell (1922) the funding provided to science by the foundations was not in proportion to the benefits they derived from science. He believed that "it may in the long run be safer and even more profitable for men of science to be free from the charity and control of the classes of privilege and sell their services to the people for what they are worth" (547).

It was important for a scientist wishing to develop programs to know the foundation personnel and to be known by them. There was a strong social network, membership in which was advantageous. Robert Yerkes gained passage, if sometimes tenuous, into this network, and it was one factor in his success in eventually getting his anthropoid station. Looking back, he wrote that "I have been in more or less intimate touch with at least some of the officers of each of these great foundations since their organization and for at least 35 years. . . . there has been established a relation which, although desirably impersonal, is the basis of a cooperative attitude and appreciation which otherwise I should not enjoy" (Yerkes 1955b).

Lobbying for a Laboratory: The Early Phase

In 1892 it took William Rainey Harper and George Ellery Hale just about an hour to secure from Charles T. Yerkes (no relation) the funds necessary to build the Yerkes Observatory (Osterbrock 1997). It would take Robert Yerkes much longer to secure funding for the Yerkes Laboratories. Biology was in transition at this time. The modest facilities and equipment that were needed by biologists at the beginning of the century were becoming inadequate (see Clarke 1987). The cost of research escalated rapidly and the problems encountered in generating funds became a matter of primary concern for laboratory directors. Yerkes' efforts are instructive and representative. Although obtaining facilities to maintain apes may be more involved than with any other nonhuman species, many of the problems he faced were also faced by others.

Recruiting Support Among Scientists

Yerkes first talked in public about an anthropoid station during the April 1913 meeting of Titchener's "Experimentalists" group at Wesleyan University (Yerkes 1914a). Important psychologists, such as Edward L. Thorndike and Edward B. Titchener, expressed enthusiasm for the plan. The idea that the station should be located in Borneo apparently came a few weeks later from William McDougall, who had lived and worked in Borneo (Anonymous, n.d.).

The next year Yerkes secured written endorsements from some leading scientists including Herbert S. Jennings and John B. Watson of Johns Hopkins, Henry H. Donaldson of the Wistar Institute, and Titchener. Watson (1916a) especially wanted a station that would combine the study of human and nonhuman subjects.

Carnegie Approaches

In May 1913, at the suggestion of John B. Watson, Yerkes wrote to Alfred G. Mayer, of the Dry Tortugas Laboratory of the CIW, informing him of his plan and seeking support "to raise a hundred thousand dollars or more in order to establish a permanent station in Borneo, in some part of Africa, or in one of the West Indies" (Yerkes 1913). By April 1914, however, Mayer (1914) informed Yerkes that funding looked improbable.

Yerkes (1916b) later tried the Carnegie Corporation, submitting a proposal for a small-scale five-year feasibility study for work at developing an anthropoid station with a budget of $25,000 the first year and $15,000 thereafter. The request was denied.

Another avenue to the Carnegie Institution was through John C. Merriam, President of the CIW, with whom Yerkes had become friendly during his years in Washington. Yerkes' diary of April 23, 1919 includes an entry that he had "had a good conference with Dr. Merriam about several matters—and among them my anthropoid station plan" (Yerkes 1919). The CIW provided support for Yerkes' 1924 trip to Cuba and for his associate, Harold Bingham's (1932) field study of gorillas. However, Merriam (1927) eventually made it clear that "it is not probable that the Carnegie Institution will be able to finance the setting up of a fully equipped anthropoid station either in this country or in the tropics in the near future."

An Approach to Scripps

In 1915 Yerkes received a letter from newspaper publisher E. W. Scripps, who was then developing the Scripps Institute for Oceanography in La Jolla, California, responding to Yerkes' (1915b) plea for a remote station. Scripps envisaged "a Psycho-Biological or rather a Bio-Psychological institution that is governed by its staff and that is entirely free from all outside influences such as might be used by politicians or so-called benefactors" (Scripps 1915, 3). Yerkes must have salivated at the thought! Exchanges of correspondence related to the proposed station among Yerkes, Scripps, and William E. Ritter, of the La Jolla facility, continued sporadically until 1926. The possibility of a site on Marco Island, Florida was considered. In the end, Scripps was unable to commit the kinds of resources required to provide primary support for the proposed station.

In Print

Although the 1916 *Science* paper (1916c) is properly regarded as the first full statement of Yerkes' plan in print, he also presented the idea in at least two earlier publications. Yerkes (1914b) concluded an article with a two-paragraph description and justification of a permanent station for the study of apes, at this point to be located in Borneo or Jamaica. Yerkes (1915b) made a similar proposal in another article, but with no mention of primates as a special interest.

In the 1916 *Science* paper Yerkes presented the most complete version of the plan to date. Among the possible locations at this point were "Borneo, Hawaii, southern California, Florida, the Panama Canal Zone, Jamaica, and the Canary Islands" (231). Research would be focused on "(1) systematic and continuous studies of important forms of behavior, of mind, and of social relations; (2) similar studies of physiological activities, normal and pathological, with adequate provision for medical research; (3) studies of heredity (genetics), life history, embryology; (4)

research in comparative anatomy, including gross anatomy, histology, neurology and pathology" (232). At this point, Yerkes believed that, once the institute was firmly established, other species might be added. Perhaps he thought that this might enhance his chances of financial support. As of 1916 the proposed institute would have six highly trained scientists in addition to a business manager, a clerical force of three individuals, a skilled "mechanician," a carpenter, and at least four laborers. It would require an endowment of about one million dollars and an annual budget, if located in California, of approximately $50,000. The importance Yerkes attached to this proposal is clear: "I am so entirely convinced of the scientific importance and human value of this kind of research that I am willing to devote my life wholly to it" (234).

Eleven years later, Yerkes still would not have his anthropoid station but he had conducted much primate research and could present a more mature plan. He elucidated thirteen "assemblages of problems" to be studied (1927a). The three-pronged organization that would later characterize the YLPB was spelled out here, with proposals for special laboratories such as those at Yale University, a breeding and observation station, and field research.

THE ROCKEFELLER FOUNDATION

On several occasions, Yerkes appealed to the influential Flexner brothers. Yerkes (1914a) explained his plan to Abraham Flexner, of the RF General Education Board, and emphasized, with some prescience, the issue of conservation writing "a few decades or generations hence, the anthropoid apes, or some of them, will probably have disappeared forever" (2–3). Yerkes (1916a) also sent his plan to Simon Flexner, Director of the Rockefeller Institute for Medical Research.

Although it was the Rockefeller Foundation (RF) that eventually funded the anthropoid station, the path to that funding was not a smooth one. Yerkes discussed his plan with Jerome D. Greene of the RF in the spring of 1914 and proposed his "Institute of Genetic Psychology" in writing to him the next year. Here, Yerkes emphasized the developmental aspect of his plan, arguing that "in order to understand the conduct or experience of the normal and pathological human adult we must know much about his individual and racial history" (Yerkes 1915c).

Yerkes seems to have been trying to write what his reader wanted to hear. When writing to Abraham Flexner at about the same time, Yerkes emphasized "the all-round *biological study of the primates*" (1915a). Greene rejected the overture, explaining to Yerkes that it was the Carnegie Institution, not the RF, that was in the business of promoting general research. Yerkes responded with an appeal to the RF routed through Abraham Flexner and received a second rejection from Greene (1915b).

Primate Research

While he was lobbying for a primate facility, Yerkes also was conducting research (as previously described in chapter 2). This may have been in part because of the intrinsic value of the problems addressed, but also served to enable Yerkes to work out some of the basic procedures for maintaining and testing primates. This aspect of Yerkes' efforts included the purchase of the Franklin Field Station in 1911, the 1915 sabbatical studying primate behavior with Hamilton in Montecito, the 1923 purchase of Chim and Panzee, the 1924 summer with the Abreu colony in Cuba, the 1925 purchase of four chimpanzees, and the work in Florida with Congo during the winters of 1927–30. The work produced a substantial body of research—especially when viewed in relation to the available knowledge at the time. However, it had even greater impact in establishing Yerkes as an experienced primate researcher. Later, Yerkes (1950, 282) would treat his early work as an apprenticeship for his work at the anthropoid station.

Yale

An integral part of the story of the founding of the anthropoid station concerns Yerkes' move to Yale University in 1924. Yerkes and Angell, who assumed the Yale presidency in 1921, had been friends for some time and Angell was well acquainted with Yerkes' plans for a primate facility. Even more than Yerkes, Angell had the connections that made it easy to interact with those in a position to help with his projects (Dewsbury 2003a). He and Yerkes shared many ideals. As worded by Haraway (1989), "Angell paradigmatically represented the elaborate interconnections of university, industry, philanthropy, and science policy in the development of material structures and ideologies of scientific management of society" (67).

Within his first year as president of Yale, Angell was hinting back and forth with Yerkes about the possibility of Yerkes moving to Yale. Yerkes realized that his chances for a primate station would be improved if he had a university connection. He had already proposed an institutional

Robert M. Yerkes' original quartet of chimpanzees: Pan, Wendy, Bill, and Dwina. From R. M. Yerkes: *Chimpanzees: A Laboratory Colony.* New Haven, CT: Yale University Press. By permission from Yale University Press.

structure that might have brought him to Johns Hopkins, but that plan was rejected (Yerkes 1921b). In 1922 Yerkes made his suggestion explicit for Angell: "Why not a school of psychology at Yale after the manner of the specialized type of combined teaching and research institution we have discussed several times" (Yerkes 1922).

Angell was interested; the possibility of bringing a psychologist with Robert Yerkes' reputation to Yale and of building such a facility in his own discipline was most attractive to the new president. However, Yale's funds were depleted and he could not see how to do it. The next year, encouraged by the interest of Abraham Flexner at the General Education Board of the RF, Angell asked Yerkes to prepare a brief statement on the proposed plan for Yale to be the headquarters of a psychobiological institute. His initial proposal called for an annual budget of $25,000, to cover salaries and other expenses. Yerkes listed three lines of research, all related to his own earlier work: (a) psychobiological research on sexual behavior in animals, with the possibility of attracting Karl Lashley, (b) work on human migration and mental measurement, and (c) field research on gorilla behavior to be organized by Yerkes and Carl Akeley (Yerkes 1923b). The Board rejected the proposal.

The two men continued to work the foundations, meeting with foundation officers in an effort to garner support. A year later, in Yerkes' next proposal, the $25,000 "Academic Center for Psycho-biological Research" had become a $40,000 Institute for Psychology (Yerkes 1924). The plan had been moved away from Yerkes' unique combination of interests to incorporate a broader range of interests with work on psychophysiological responses under Raymond Dodge, comparative and genetic psychology under Yerkes, behavior and the nervous system under Lashley, mental measurement and statistics under Carl Brigham, and the history and theory of psychological conceptions under Edwin G. Boring. Yerkes proposed that an institute for mental hygiene and psychiatry be associated with the facility. The 1924 proposal included a prominent section on the need for a primate station, but it was made clear that this would be independent of the Institute.

Eventually, Angell secured funding for the Institute. This can be seen, at least in part, as an outgrowth of personal relationships that had been developed earlier. Beardsley Ruml, for example, with his earlier history with Angell and who still addressed him as "Chief," was critical in this effort (Morawski 1986).

In the fall of 1924, Robert Yerkes, at age 48 and with a salary of $7,000, moved to Yale into a position that would enable him to embark on what was arguably the mature part of his career. However, the personnel attracted to the new institute fell short of his proposal. Anthropologist Clark Wissler, Raymond Dodge, and Roswell Angier, were Yerkes' initial associates at the Institute; Arnold Gesell joined them later. As usually happens in such institutes, the amount of coöperative research actually conducted fell far short of expectations.

The Institute of Psychology lasted just five years; it was superseded by an even more grandiose cooperative structure, the Yale Institute of Human Relations (IHR). This incarnation, opened in 1929, was designed primarily by Robert M. Hutchins, dean of the law school, and Milton Winternitz, dean of the medical school. This institute would incorporate a wide range of human sciences, with psychology as a core, as relevant to the practice of law and medicine (Morawski 1986). Yerkes opposed the formation of the IHR, but agreed to be a participant if it was formed. The RF granted $2 million for a building and an annual budget of $150,000.

All of the faculty from the Institute of Psychology joined the new institute, which had twenty-one faculty members altogether. However, the IHR was no more successful in fostering true integration than had been the Institute of Psychology.

Chimpanzees in New Haven

Yerkes' laboratory within the Institute of Psychology was quite active, with studies of cats by Donald K. Adams, guinea pigs by Chauncey M. Louttit, chimpanzees by Harold C. Bingham and Yerkes, and with Robert and Ada Yerkes doing the literature survey that would become *The Great Apes* (Yerkes and Yerkes 1929), in addition to other projects (see Yerkes 1927c). However, Yerkes was looking ahead. He recalled, "no sooner were we established in New Haven than I began to look about for ways of realizing this dream [of a primate facility]" (Yerkes 1950, 266).

GETTING ROCKEFELLER SUPPORT

In January 1925, a few months after he arrived in New Haven, Yerkes sent a proposal for primate research at Yale to John C. Merriam at the CIW (Yerkes 1925b). He asked for an annual budget of $10,000. Yerkes sent a copy of the proposal to Edwin Embree at the RF, with whom he had been in conversation, who circulated the proposal to other Rockefeller officers with a positive recommendation (Embree 1925). Apparently, as a result of further consultation, Yerkes submitted a revised proposal to Embree in April (1925c). This proposal was more concrete, calling for four years of funding to begin on July 1, 1925 with an annual budget of $10,000 from the Rockefeller Foundation to be used together with monies from Yale and the Institute. Emphasis would be placed on "the study of important aspects of ideation and sex behavior and on preparation for the experimental study of genetic problems." The work would be done in New Haven, with the ultimate goal of a primate breeding station in a more hospitable climate. H. C. Bingham, who had been with Yerkes in Havana, would move from Wesleyan University to become the Research Associate. One senses that Yerkes knew that he was getting close.

Yerkes laid out for Embree his two justifications for ape research (1925d). On the one hand, he pointed out, there are many problems that simply cannot be addressed in humans for ethical and other practical reasons. He suggested that the Board treat the use of infrahuman animals in psychobiology the same way in which they would in medical experimentation. In his second tack, Yerkes pointed to the importance of understanding the genesis of behavior—its phylogenetic origins and its ontogeny. These were presented first as important problems in and of themselves, and independent of generalizability to humans. As might be expected, however, he added that "we could scarcely act more stupidly or shortsightedly in our efforts to gain adequate knowledge of human behavior than by limiting our interest and effort strictly to humanity" (1925d).

The Rockefeller Foundation Board met on May 27, while the Scopes trial in Tennessee was near its peak, and voted the four-year, $10,000 annual budget for Yerkes' project (Thompson 1925). He did not yet have his breeding station, but after the many years of trying he finally had foundation support for ape research.

THE MANSON BARN

Yale made available to Yerkes a vacant brick building, the "Manson barn" at 135 Prospect Street in New Haven. It would serve for six years. When the Yerkes family and Bingham returned from Franklin in September 1925 they brought with them Bill, Dwina, Pan, and Wendy, and the chimpanzee facility at Yale was in business. Work done there showed that it was indeed feasible to work with higher primates on a long-term basis. Yerkes gave Bingham much of the credit for working out maintenance procedures. Otto Tinklepaugh and Henry Nissen joined the staff during these years.

The results of the first two years of research, perhaps not surprisingly, were not overly impressive (Yerkes 1926d, 1927e). Problems of development, sexual behavior, learning, and cognition were explored. A highlight was Bingham's (1929b) study of box stacking using traditional methods, following upon work done by others. Bingham believed that he saw evidence of ideation. Tinklepaugh (1932a) studied delayed-response learning. The work in these studies suggested that the performance of chimpanzees in various tasks differed from other species and thus might help justify expenditure of the additional funds to breed and maintain them, rather than other nonhuman primates.

Along the Road to Orange Park

As important as the Yale facility was, it was but one important step on the path to what Yerkes really wanted. He continued to lobby toward that end. President Vincent of the Rockefeller Foundation visited the Yale operation in April 1926 and conferences with Vincent, Embree, Vernon Kellogg of the National Research Council, and Merriam continued. What is striking from a contemporary point of view is how much the decision regarding the station depended upon the judgments of the foundation officers rather than on peer review, which would become standard later in the century.

Securing an Initial Commitment for Funds for a Station

Establishment of the primate laboratory at Yale only served to accelerate Yerkes' efforts to secure a true anthropoid station. He tried many tactics in his efforts to wrench free the necessary funds. At one point, obviously frustrated, he appeared to try to threaten Embree at the RF with abandonment of the whole project: "I have definitely decided that if within a few months I cannot secure proper provision for the continuance and extension of anthropoid research, I shall abandon it in favor of an important task [unspecified] which I necessarily dropped when in 1917 I left Harvard for war service" (Yerkes 1926b). Of course, this had just the opposite effect as Embree (1926a) questioned his commitment to the project. Yerkes (1926c) rejected Embree's suggestion that some other facility might breed chimpanzees, freeing Yerkes for research. Yerkes stressed the advantages of a breeding facility for securing complete life histories of animals born therein.

Embree was becoming more receptive to the idea of the station. Yerkes was unaware that Embree was testing the waters for support. Embree wrote to Simon Flexner indicating that "I am inclined to think that the support of such a breeding station would be desirable" (1926c). Flexner was not supportive. By November, however, Embree (1926b) would have to inform Yerkes that funding for the breeding station would not be forthcoming—at least not for the present. Yerkes (1926b) informed Embree that he planned to appeal directly to Mr. Rockefeller! The man was nothing if not persistent. The officers expressed no objection to Yerkes' latest tactic, although Yerkes did receive a letter from George Vincent (1926), the president of the Foundation, explaining that the policies of the RF were under a review that forced caution in making lasting commitments.

Events in 1927 continued the pattern established in 1926, with Yerkes contacting various foundation officers, the officers communicating among themselves, and no commitment for an anthropoid station forthcoming. With the reorganization of the RF in 1928 responsibilities were fluid and hard to follow. Responsibility for the funding of the Yale primate facility was switched to the new Division of Life Sciences, under Max Mason (Pearce 1928). Yerkes' efforts had been caught up in the transition. By the time he assumed office, Mason found several letters from Yerkes already awaiting him; Mason's response was to write to Angell asking if he supported Yerkes' project (Mason 1928a). Angell (1928) provided unqualified support for both Yerkes and the project, noting that the anthropoid station was a project with which he had "had active and interested contact for nearly a decade." Mason visited New Haven and conferred at some length with Yerkes on October 12, 1928 (Mason, 1928b).

Although he did not know it, Yerkes was about three months away from initial funding for the anthropoid station. The historical record for this critical period, however, seems incomplete. It appears as though Mason asked Yerkes for a revised proposal, which Yerkes promptly forwarded (Yerkes 1928b). This proposal resembled the earlier ones quite closely, specifying several lines of research, including studies of the chimpanzee estrous cycle, growth and development, genetics, social behavior, the nervous system, psychopathology, etc. The site would be twenty to fifty acres with three buildings, living quarters, support space, and laboratory space, and approximately twenty animals. Yerkes anticipated expansion as the project progressed. The personnel would include two investigators and a variety of support staff. Yerkes estimated that to continue the New Haven operation and build and maintain the station for five years he would need a total budget of $325,000.

Several things happened to the proposal at this point. For one thing, the proposal was referred back to Lawrence Frank and the Laura Spelman Rockefeller Memorial because the Foundation was unable to take action quickly enough (Frank 1928). There appear to have been several meetings regarding the kind of control that would be necessary were the new facility to be established. Yerkes (1928c) sent the expanded materials to Frank. Much of this material was more legalistic in establishing control. A Scientific Advisory Board of four to seven members would be established to advise the director on scientific matters. A division of responsibilities in the event of a decision to separate the anthropoid research station from Yale University was spelled out. The budget was refined.

The executive committee of the RF considered the proposal in January of 1929 and recommended support, but contingent upon a feasibility study to be completed before major funding was provided:

> RESOLVED that the sum of Twenty-five thousand dollars ($25,000) be, and it is hereby, appropriated, of which so much as may be necessary shall be paid to

> Yale University to defray the expenses of a study by a committee of experts of the feasibility of establishing an anthropoid breeding station; with the understanding that upon receipt of a favorable report and a recommendation from this committee, the Foundation will be willing to consider further contribution of not more that $475,000 toward the establishment of such a station and the support of it and of a program of anthropoid research over a period of years. (Thompson 1929a)

Another Committee

Yerkes still did not have his anthropoid station. An elite committee would have to be established and persuaded to make a positive recommendation to the Rockefeller officers. The committee was established with Edwin G. Conklin of Princeton University, Milton J. Greenman of the Wistar Institute of Anatomy and Biology in Philadelphia, John C. Merriam, the president of the CIW, Theobold Smith of the Rockefeller Institute for Medical Research, H. Gideon Wells of the S. A. Sprague Memorial Institute, and Clark Wissler of the American Museum of Natural History. Angell chaired the committee and Yerkes acted as secretary. The committee met in New Haven on February 17, 1929; all but Merriam attended.

Prior to the meeting, Yerkes (1929b) prepared an extensive, twenty-eight-page memorandum in which he made his case for the station. The Rockefeller officers were concerned about the care and maintenance of great apes in captivity. There were many questions. Can anthropoids be kept for long periods in captivity? Will they be in normal condition? How serious are disease risks? Can nutritional problems and dietary requirements be worked out? Will the animals breed in captivity? Will the conditions of captivity ensure social and psychological normality? Can animals be tested and studied beyond childhood? Will the colony be an adequate source of laboratory animals? Will the animals stand operative treatment? Yerkes prepared a response, relying on his experience in New Haven and the work at Kindia, Tenerife, Moscow, and Havana. He included letters testifying to the feasibility of work with captive chimpanzees from authorities experienced working with them. The committee approved the plans and asked Yerkes to work further to find an appropriate site within the boundaries of the United States (Minutes, 1929).

Finding a Site

Throughout his long campaign for the facility, Yerkes considered appropriate locations for an anthropoid station. Eventually, he decided that it would be more practical to bring the apes nearer to him than to establish the facility near the apes. Settled at Yale, it seemed that Florida offered the most promise for both adequate climate and accessibility. Although the land boom collapsed in 1926, Florida was now more prominent in the public mind than previously. By the time Yerkes was ready to purchase property, the price already had gone down.

Yerkes began seriously surveying the state searching for a site during the three winters that he studied Congo during 1926–28. Hurricane risk, high land prices, and inconvenience of access eliminated many locations. In February and March 1928 Yerkes visited a number of sites throughout the state of Florida; Bingham also visited all except one. They considered the area around Sarasota in the west. A trip was made to Vero Beach and Miami along the Atlantic coast. They went to such places as Winter Haven, Haines City, Sanford, and Bartow in the central lake region. They spent March 5 and 6 in Gainesville considering the advantages of the university city, the site of the University of Florida (Yerkes 1929c). They considered the Jacksonville area, with which Yerkes was quite familiar from his work with Congo. Yerkes then developed an elaborate rating scale for different criteria related to meteorological, geographical, geological, botanical, zoological, economical, sociological, and psychological factors. Using this scale, Yerkes concluded that the Jacksonville and Gainesville areas offered the most promise. The strength of Jacksonville was its accessibility; that of Gainesville, the university community. The isolation from urban centers and cultural activities were strong factors working against the central lake region.

With the committee's mandate, the search for a site was resumed, with Greenman, from the Advisory Committee, accompanying Yerkes in February and March 1929. At this point, Lake Wales had been restored as a possibility and it, Gainesville, and Jacksonville were under consideration. In the end, however, Orange Park, near Jacksonville, won out. It had the advantages of accessibility, with the newly built roads and the Atlantic Coast Railroad, and the availability of a good site at a good price. They could buy 200 acres of high and dry land for just $50 per acre. Price may have been the determinative consideration.

Securing the Final Commitment

On March 20, 1929 Angell (1929) collected the results of the committee meeting and Yerkes' reports on possible sites and reported to George Vincent, President of the Rockefeller Foundation, that all was in order for the establishment of an anthropoid station in Orange Park, Clay County, near Jacksonville, Florida.

Two days later the Rockefeller Foundation committee met and passed the resolution that finally established Yerkes' anthropoid station:

> RESOLVED that the sum of Four hundred seventy-five thousand dollars ($475,000) be, and it is hereby, appropriated, of which (1) $90,000 or as much thereof as may be necessary shall be paid to Yale University for preliminary reconnaissance, administrative expenses, development of plans purchase of site, construction of plant and initial supply of animals in connection with the establishment of an anthropoid breeding station; and (2) $385,000, or so much thereof as may be necessary over a ten-year period beginning July 1, 1929, shall be paid to Yale University to provide for a program of anthropoid research, with the understanding that not more than $25,000 shall be paid during the year beginning July 1, 1929, and ending June 30, 1930, and not more than $40,000 in any one of the succeeding 9 years. (Thompson 1929b)

The job of getting the station was finally complete. The job of building the station was about to begin. Yerkes would recall: "in 1929 the scene of our labors was shifted from the heart of Yale to the more open spaces of Florida and from the somewhat rigorous climate of New England to the 'sunshine and ocean breeze' of north Florida" (Yerkes 1950, 286).

Robert Yerkes was not known for his sense of humor. However, the Rockefeller Foundation soon received a memorandum from the four chimpanzees, William, Dwina, Pan, and Wendy, expressing a wish to return to Africa (Yale Anthropoids, 1929). Vincent (1929) replied that he could not help; "whenever we give money to an institution we assume an almost simian air of detachment." Vincent added that as he was retiring at the end of the year he had hoped that the chimpanzees might come to New York and assume control of the Rockefeller Foundation.

One senses that the Rockefeller officers retained some healthy skepticism regarding the anthropoid project. Embree (1929), who had left the Rockefeller Foundation, wrote to Mason expressing delight that the little appropriation he had made to get the Yale facility started had blossomed into one for a full primate station. Mason replied that "we went into such measure only after a very careful study by everybody whose field was touched by this work. I think it was somewhat chancy, but large values may come" (Mason 1929). Clearly, Yerkes was still on probation.

Orange Park

The unique flavor of the Yerkes Laboratories was shaped by many forces, some of which came from the Florida environment in which it was estab-

Photograph of Robert M. Yerkes, the first director, circa 1930. Courtesy of the Archives for the History of American Psychology, University of Akron, Akron, OH.

lished. To understand the essence of the laboratories, one must know something of the geography, historical context, and general way of living within which they existed. Ada Yerkes understood this well and went so far as to write an eighty-nine-page monograph on the history of the area, "Yale in Florida" (A. W. Yerkes 1939).

The Town of Orange Park is in the northeastern corner of Clay County in northeastern Florida. It lies on the west bank of the St. Johns River. The southern boundary of Orange Park is formed by Doctors Lake, actually a wide inlet of the St. Johns River that connects to the St. Johns. According to the 2000 census, the population of the town at that time was 9,081. This is deceptive, however, because much of "Orange Park," including both the site of the YLPB and the present Orange Park High School, lies outside of the town limits. Orange Park today is part of a metropolitan area, a bedroom community of Jacksonville.

The major east-west road is Kingsley Avenue, on which the YLPB was located. The six-lane roads, modern shopping mall, and other amenities make the Orange Park area a prime example of American life in the early twenty-first century. That growth is recent, however, and the Orange Park in which Robert Yerkes founded the YLPB was a small southern town.

The early history of much of the region entails conflicts among native Americans, French, Spanish, English, pirates, and eventually United States nationals (see A. W. Yerkes 1939; Blakey 1976). In 1803 the lands of Orange Park were purchased by Zephaniah Kingsley, Jr., a naturalized U. S. citizen of British ancestry and the man after whom the road on which the YLPB would be located was named. In the words of Bennett: "he was a slaveholder whom his neighbors branded an abolitionist because of his writings and speeches on the civil rights of blacks" (1989, 89).

The land passed through the hands of several owners and Orange Park was developed. Timber, cattle ranches, phosphate mines, sugar mills, and turpentine stills were begun. The Civil War had a devastating effect on the area. Later, as tourism and citrus production blossomed, the towns along the river thrived; both steamboats and oranges had a major impact on growth. The town of Orange Park was incorporated in 1879.

The fortunes of Orange Park ebbed and flowed throughout the twentieth century. By 1939 Ada Yerkes reported that "the roadways are perfect," but noted that "pigs and cattle browse, a menace to the unwary motorist since they may decide at any moment leisurely to cross the road" (5). Electricity arrived in 1924. In 1928 the town still had just 170 registered voters (Adams 1994) and a population of 661. Perhaps the most notable institution in Orange Park was Moosehaven, a retirement community operated for its members by the fraternal organization, the Loyal Order of Moose, that began during the 1920s and continues today.

World War II had a major impact due to the proximity of the Jacksonville Naval Air Station to the north, Camp Blanding to the southwest, and a naval complex in nearby Green Cove Springs.

The development of Orange Park as a modern residential center and "bedroom community" for the City of Jacksonville began in earnest in 1946 with the development of Holly Point, which fronts both the St. Johns River and Doctors Lake (Blakey 1976).

The Press

Establishment of the anthropoid station was first formally announced in a press release from Yale on October 30, 1929. It and many press clippings were collected in the YLPB scrapbook. The press release provided basic information about the new station, the Rockefeller grant, the membership of the advisory committee, the purpose of the research, the tripartite organization of the laboratory, and Robert Yerkes himself. In actuality, however, Angell jumped the gun by announcing the anthropoid station in public at the Ninth International Congress of Psychology that met in New Haven in September.

In general, the press gave the new station fairly straightforward coverage, although a few writers seized the opportunity for efforts at humor. The *Tampa Tribune* of October 18 asked "If an ape apes a Jacksonville man is the ape apt to evolve?" The *Monticello News* of October 18 reprinted a quotation: "Trouble is we are all afraid they might want us to turn them loose and let them vote about three generations later" (Scrapbook, n.d.).

The main flak came from the newspapers in the Tampa Bay area. The *St. Petersburg Independent* of October 31 ran a story suggesting that the station should be located in Sarasota, if the city wanted it, rather than in Orange Park because of the proximity to the Ringling operation (Scrapbook).

The *Tampa Tribune* created the most fuss in complaining about this "monkey business" in an article entitled "A Half-Million for Apes." "We think much more valuable results might be obtained from a half-million dollar study of certain varieties of the human species, with a view either to improving or exterminating them." The story does not specify in print which variety of the human species would be exterminated. Yerkes wrote a reply that was published on November 18. He repeated his belief that the research would be of practical utility to humans and distanced himself from studies of evolution stating that "what we are proposing to do is not inspired by interest in the problems of human evolution." This was, after all, a southern newspaper in 1929. The *Tribune* took another swipe at the project the next spring. For the duration of his administration, Yerkes would take a low-profile approach with the station. Such press coverage must have been one factor in making him gun-shy of too much publicity.

Actualizing the Plan

Administrative Matters

As all of this was happening, Yerkes had to worry about shifting administrative patterns as well as matters of scientific priorities. As the RF was being reorganized, the Institute of Psychology was metamorphosing into the IHR. Yerkes was caught in administrative reorganization at both ends of his academic and administrative worlds. Where to put the new tripartite anthropoid program at Yale? In 1929, Yerkes wanted to keep it as a separate administrative department within the new Institute of Human Relations (Yerkes 1929e). However, he felt that the facilities in both the Manson barn and the new IHR building would be inadequate for the housing of chimpanzees and thus wanted his New Haven operation included in new construction being undertaken as an extension in the west wing of the Sterling Hall of Medicine. This would have the advantage of proximity to both the IHR and the Department of Physiology and was where the animals would eventually be housed (Yerkes 1929f).

The group of advisors that had met in February became the Advisory Board for the Anthropoid Experiment Station of Yale University at Orange Park, Florida, the first official name of the YLPB. They held their first meeting as that board in New Haven on December 7, 1929. Among the recommendations to emerge was that "the subtropical experiment station be closed to all except professional visitors or those officially connected with the organization, and that all visitors be required to wear masks, gloves, and gown for protection of chimpanzees and themselves" (Minutes of Advisory Board for the Anthropoid Station December 7, 1929). The pattern of exclusion was probably necessary for undisturbed scientific work, but, carried to extremes, would turn into a problem for Yerkes some years later.

To Africa

Yerkes wanted to visit the other primate facilities of the world before establishing his own. On June 1, 1929, Robert and Roberta Yerkes sailed for France and then went to Berlin, Moscow, and Leningrad. In Moscow they visited with Natalie Ladygin-Kohts, who had been working for some years with her young chimpanzee, Ioni. The laboratory was in the Darwinian Museum, which her husband had founded, and had many unpublished photographs and documents. After some European touring with the rest of the family. Robert and Roberta then sailed from Marseilles to Conakry and Kindia to visit the primate station there and arranged for the transfer of some chimpanzees to help stock the new station, and for a member of the Yale staff to travel to Kindia to conduct a study of chimpanzees under field conditions (Yerkes 1929d). Yerkes then returned home to face the task of making his dream a reality to fruition. It had been a long road, but Robert Yerkes was finally ready to start running, rather than dreaming about and planning for, an anthropoid facility.

4

Running a Primate Station

In some respects, the year 1930 was an inauspicious time for the opening of a new laboratory. On "Black Friday," October 28, 1929, the United States stock market collapsed, triggering economic depression and a worldwide financial crisis. This would have major consequences for many individuals and institutions.

On the political scene, Herbert Hoover was in the second year of his term as President of the United States. Culturally, popular songs of the day were "Time on My Hands," "Body and Soul," and "Walkin' My Baby Back Home." Moviegoers saw Marlene Dietrich in *Blue Angel* and *All Quiet on the Western Front* won an Academy Award. Sinclair Lewis won a Nobel Prize for Literature for *Babbit* and Sigmund Freud published *Civilization and its Discontents* (Grun 1991).

Opening the Laboratory

Meanwhile, Robert Yerkes' primate station opened with little fanfare but much activity. The Orange Park property was purchased through the offices of Charles A. Brown, a Realtor in Jacksonville. The plot of nearly 187.56 acres was approximately a mile long and one-third of a mile wide, about a mile from the village of Orange Park (Annual Report 1930; A. W. Yerkes 1939). The land lay approximately sixty feet above sea level on a "blackjack ridge," called that because of its scrub oaks, pines, and live oaks. Water was supplied by an artesian well. Plans for the buildings were prepared by Richard A. Kimball and Ellery S. Husted, working in the offices of James Gamble Rogers in New York. The resident supervising architect was W. Kenyon Drake; the general contractor was Arthur Perry, Jr., both of Jacksonville (A. W. Yerkes 1939).

Ground was broken in February. Construction of the initial buildings of the anthropoid station was completed in June 1930. The station was designed to house about twenty-five chimpanzees. There were three main units. The Laboratory, built of hollow tile and stucco, had twelve rooms that served as offices and experimental rooms. A Service Building housed a kitchen, a storeroom, and a shop equipped with lathes, saws, and other tools; heating equipment was in the basement. The Quarters Building, made of reinforced concrete, steel, and heavy metal wire, housed the adult animals. It included eight animal rooms with adjacent outdoor cages and two small service rooms. Apparatus to control lighting, ventilation, and the operation of cage doors was on the second floor. The Laboratory was in the front of the facility, the Quarters Building in the back, and the Service Building in between. A covered corridor connected the buildings. Smaller buildings within the fence included a pump house, Hospital Building, and Storage Building. Outside the fence lay a garage and an old frame building used for storage.

The one house that remained on the property was remodeled as a home for laboratory superintendent William C. Atwater. All of the buildings, except the Superintendent's house, were surrounded by and enclosed within a seven-foot-high fence of chain-link netting.

Throughout his career, Yerkes seemed fascinated with buildings and gadgets. He seemed never to tire of publishing descriptions, floor plans, and photographs of the facilities within his domain (e.g., Yerkes 1932a; 1932b; 1943). Perhaps such planning and organization, rather than conducting of the actual research, were his real source of joy.

When the nine buildings finally were ready, Yerkes and his staff wasted little time in getting started. After inspections, Richard A. Kimball, representing the architects, formally accepted the buildings on June 9, 1930; staff member Otto L. Tinklepaugh arrived with the first four chimpanzees, Bill, Dwina, Pan, and Wendy, the next day (Annual Report 1930). After all of the media coverage surrounding the announcement of plans for the laboratory, there appears to have been no press coverage of the actual opening. That is probably how Robert Yerkes wanted it. Just seven days later, Henry Nissen arrived from French Guinea with sixteen additional animals from Kindia on board. Yerkes met Nissen at the ship and the two brought six young animals to the New Haven facility and ten older animals to Orange Park. There were now fourteen chimpanzees in the Orange Park colony, among which two males and two females were sexually mature. Dwina was already pregnant when she arrived in Orange Park. The proud scientists could announce the birth of Alpha, daughter of Dwina and Pan, on September 11, 1930. The Anthropoid Station of Yale University was up and running.

William C. Atwater reported for duty as the first superintendent of the station on June 1. He had formerly worked in the Treasurer's office at

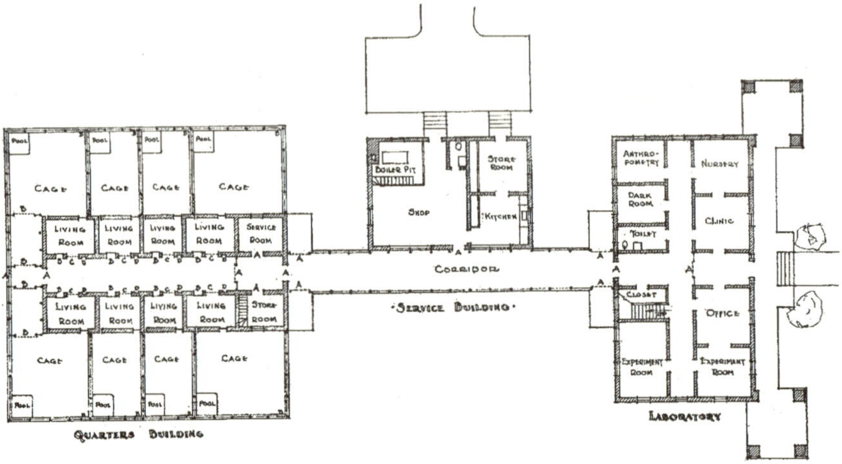

Floor plan of the original buildings of the Yale Laboratories of Primate Biology. From R. M. Yerkes: *Chimpanzees: A Laboratory Colony.* New Haven, CT: Yale University Press. By permission from Yale University Press.

Photograph of the original Laboratory Building, later known as the Administration Building, of the Yale Laboratories of Primate Biology. Courtesy of the American Philosophical Society, Philadelphia.

Yale. Robert and David Yerkes, Mr. Atwater, and Otto Tinklepaugh spent most of June working out the kinks, and taming and caring for the animals. They could report that the physical plant was in reasonably good shape by July 1, although the animals still had some adapting to do (Annual Report 1930).

An Overview of the Yerkes Years (1930–1941)

Robert Yerkes would enjoy a tenure of just eleven years as the director of the facility that he had worked so hard to establish. During that period, much of Yerkes' time would be occupied with details of maintaining and administering the laboratory, thus limiting his hands-on research. Yerkes relied on the personnel he was able to bring to Orange Park to conduct most of the actual experimentation. The personnel problems he faced were difficult, however. He had a hard time attracting and keeping good people; handling the personnel once they arrived was not his great strength. During the period of Yerkes' tenure, the laboratories would undergo several administrative changes that affected the way in which they were run. Yerkes would make repeated efforts, some of them successful, to improve the physical plant and to expand the operation. Funding and finances would be perennial problems. The practical problems of running a facility for ape research, especially the care and breeding of the animals, would basically be solved, an important accomplishment, but the output of research during the Yerkes years would be judged by some as disappointing.

Developments in Orange Park, at the RF, at Yale, and elsewhere would be such that by 1937, when deliberations about a renewal of support for the Station began, Yerkes would be in conflict with the officers of the Rockefeller Foundation and it would become clear that the only way to ensure the future of the research station was for Yerkes to agree to withdraw from its management. Thus, Yerkes' role in the administration of the YLPB can be seen as decreasing during the years from 1937 until his retirement as Director in 1941. Clearly, Yerkes was a better planner and developer than he was a day-to-day administrator. That caused many problems; these matters will be explored in turn.

Administrative Structure

The Orange Park facility opened in 1930 as the "Anthropoid Experiment Station of Yale University"; on July 1, 1931, following the discontinuation of the Yale Institute of Psychology, it was reorganized as the "Laboratories of Comparative Psychobiology" as a section of the Department of Physiology, chaired by John F. Fulton, in the Yale School of Medicine. In August 1931, the New Haven laboratory was moved from the Manson Barn to Wing B of the Sterling Hall of Medicine. After three more years, the Laboratories were removed from the administrative unit of the Department of Physiology and given an independent status in the university group of physiological sciences.

Establishing a Three-Pronged System

Yerkes had, for some time, envisaged a tripartite structure for the primate laboratories, including the laboratory in New Haven, the Orange Park station, and field sites; it was in 1932 that the organizational structure became formalized and annual reports were organized according to these three designations.

The main functions of the New Haven facility were to do those kinds of research done best as cooperative ventures with Yale colleagues, serve as a center of instruction, and be a permanent headquarters, where staff from the other sections could complete those aspects of their work dependent on a strong library. The Orange Park facility allowed the breeding and rearing of chimpanzees and the study of those problems better addressed in the subtropical environment. Field research was critical for understanding the animals in their natural habitats. Yerkes was a believer in the importance of field research, but believed that it had to be conducted opportunistically.

The tripartite organizational structure would provide Yerkes' primary organizational foundation for the first few years of the station. Yerkes was unable to sustain the field projects, however, and this third part of the plan soon became a hollow sector. Yerkes was especially disappointed that he was unable to develop field studies of bonobos; his interest is stated repeatedly in documents of the time.

By 1936, the "northern division" and "southern division" were still keys to Yerkes' thinking, but, as reflected in the annual reports, by then Yerkes was thinking in terms of a more unified structure and the divisions became subordinated to the overall unified program as he looked back each year.

Delegating Some Responsibility

For the first ten years of the Station, the Yerkes family split its time among three sites in an annual migratory pattern. They would be in Orange Park from February through July each year, go to New Hampshire for four to five weeks after that, and then return to New Haven after the meeting of the American Psychological Association in early September. The pattern would end on July 26, 1939, the last summer of the Yerkes family in residence in Florida (A. W. Yerkes 1957). During 1940, Yerkes would make only three visits to Orange Park.

Clearly, Yerkes had to delegate authority in New Haven and Orange Park when he was away from either site. Henry Nissen assumed the responsibilities as Deputy Director in New Haven from the founding of the Orange Park facility through 1939; he became the Assistant Director in charge of work in Orange Park in 1939 when Yerkes terminated his pattern of residence in the south. The responsibilities of a Deputy Director in Orange Park fell on Otto Tinklepaugh during 1931–1933 and on James Elder until succeeded by Nissen.

Scientific Advisory Board

The original Advisory Board remained intact until 1935, when Smith died and was replaced by Frank R. Lillie of the University of Chicago. The Board often met in March in Orange Park (A. W. Yerkes 1939).

The Administrative Committee of the Yale Laboratories of Primate Biology

As best as can be determined, the Scientific Advisory Board did not meet after 1937. With the reorganization that occurred as part of the "crisis of 1937" (see chapter 7), it was replaced by an Administrative Committee of the Yale Laboratories of Primate Biology. This Committee had more power to make appointments and other important decisions affecting the YLPB than had the original Board. The members of the initial committee were Carl G. Hartman of Johns Hopkins University as chairman, Edgar Allen of Yale, William H. Taliaferro of the University of Chicago, Leonard Carmichael of the University of Rochester, and Yerkes, appointed with a set of staggered terms. The Dean of the Yale School of Medicine served as a member *ex officio*.

This group first met in New Haven in January 1939. The committee approved several actions that would allow William C. Young, then of Brown University, and his associates to move their program from Providence to Yale. That action meant that his whole program, including significant amounts of research on rats and guinea pigs, would come under the auspices of the YLPB. With its new power and responsibility, the Committee came to consider and evaluate the initiation or continuation of specific lines of research. Earlier this had been largely the province of the director.

By the time of the second meeting, it was apparent that power had shifted further from the director to the committee. Chairman Hartman argued for increased use of monkeys, as opposed to apes, and the use of the facilities by more visiting investigators (Minutes of Administrative Committee Meeting December, 1939). At the third meeting, the first in Orange Park, the power of the committee became even more apparent. Now, the committee was voting approval or disapproval of specific research programs proposed by investigators at the laboratories (Minutes of Administrative Committee Meeting, March 1940).

As the new dean, Francis G. Blake replaced Dean Bayne-Jones at the December 1940 meeting. Leonard Carmichael succeeded Hartman as Chair and Nissen was added as an *ex officio* member in 1941. The work of the committee continued, with much effort devoted toward finding a new director of the laboratories.

The Scientific Staff

A major concern of Yerkes' throughout most of his tenure as director was with the personnel that staffed the YLPB. Certainly, the work of such a facility could be no better than the people who staffed it. Yerkes' occasional periods of optimism regarding the development of his scientific staff were rare compared to his repeated complaints about its quality.

Evaluation of the staff was complicated by the difficulties Yerkes had in dealing with them. He tended toward an authoritarian management style patterned after the hierarchical organization of military establishments. There was a pattern for some, but not all, staff members in which Yerkes appeared generally supportive of them for a period of time but became critical of them around the time at which they began to assert their independence and eventually leave the YLPB. It is sometimes difficult to determine whether this was due to the emergence of real problems or, as was frequently asserted, Yerkes tended to be intolerant of those who began to think on their own and depart from his way of doing things.

Personnel on the Scientific Staff

During much of his tenure, Yerkes regarded the New Haven and Orange Park operations as the "northern division" and "southern division" of a single unit. It is thus sometimes difficult to separate the two in order to

Photograph of researchers at the Yale Laboratories of Primate Biology circa 1939: George Haslerud, Meredith Crawford, James H. Elder, Henry W. Nissen, Glen Finch. Courtesy of Jane Elder Wulff.

concentrate on Orange Park. Yerkes believed it unwise for staff to spend the whole year in Orange Park, away from the intellectual stimulation, libraries, and other facilities available in New Haven, so many of the staff, including Yerkes, would migrate between the two divisions. Yale University in the 1930s, its golden years, had one of the finest concentrations of psychologists ever assembled, with a group of outstanding students gathered about such senior faculty as Clark Hull, Raymond Dodge, Walter Miles, and Arnold Gesell. Approximately twenty-three individuals served on the scientific staff in Orange Park during these years, together with another twenty who worked only in New Haven (see Table 4.1). Nine members of the scientific staff earned Yale PhDs during Yerkes' tenure.

Comings and Goings

There was an ebb and flow of staff members; Yerkes never could put together and keep the kind of staff he wanted. He commented on the

Table 4.1.
Members of the Scientific Staff During the Yerkes Years with Birth Date, Birthplace, PhD Information, and Approximate Years of Service[1]

Harold Clyde Bingham. b. January 21, 1888, Rowan, IA. PhD (Psychology) Johns Hopkins University, 1923. "Visual perception in the chick." New Haven 1930.

John Merle Bruhn. b. October 10, 1904, Flandreau, SD. PhD Yale University, 1933. "The respiratory metabolism of primates." Orange Park 1933–1937, New Haven 1936–1937.

Clarence Ray Carpenter. November 28, 1905, Lincoln Co., NC. PhD (Psychology) Stanford University, 1932. "The effect of complete and incomplete gonadectomy on the behavior and morphological characters of the male pigeon." New Haven 1931–1934.

John Todd Cowles. b. July 31, 1911, Burlington, VT. PhD (Psychology) Yale University, 1937. "Food-tokens as incentives for learning by chimpanzees." New Haven 1934–1937.

Meredith Pullen Crawford. b. October 10, 1910, Sweetbriar, VA. PhD Columbia University, 1935. "Brightness discrimination in the rhesus monkey." Orange Park 1937–1939, New Haven 1934–1937.

Guy Hamilton Crook. b. February 7, 1909, Ravenswood, WV. PhD (Psychology) University of Minnesota, 1939. "The effect of exercise on the recovery of motor function in the rat." Orange Park 1938–1939.

James Harlan Elder. b. September 24, 1902, Marshall, IN. PhD (Psychology) Yale University, 1933. "Auditory acuity of the chimpanzee." Orange Park 1933–1939, New Haven, 1932–1939.

John Lincoln Finan. b. December 27, 1911. PhD (Psychology) Yale University, 1937. "The effect of frontal lobe lesions on temporal discrimination in the monkey." New Haven 1937.

Glen Finch. b. October 23, 1909, Portland, OR. PhD (Psychology) Yale University, 1935. "The effect of protracted exposure to intense tonal stimuli on auditory acuity in dogs." Orange Park 1937–1942, New Haven, 1937–1939.

Walter Frank Grether. b. March 26, 1912, Stewart, WI. PhD University of Wisconsin, 1938. "Analysis of color vision in New-World and Old-World monkeys." New Haven 1938–1941.

Ross A. Harrison. b. March 7, 1912, Elizabeth City, NC. PhD (Psychology) Johns Hopkins University, 1938. "Personal tempo and the interrelationships of voluntary and maximal rates of movement." Orange Park 1939–1940.

continued

Table 4.1. *Continued*

George Martin Haslerud. b. September 14, 1906, Fargo, ND. PhD (Psychology) University of Minnesota, 1934. "A comparison of free and prescribed maze patterns of the rat." Orange Park 1934–1935 and later summers as visiting investigator, New Haven 1934–1935.

Ernest Ropiequet Hilgard. b. July 25, 1904, Belleville, IL. PhD (Psychology) Yale University, 1930. "Conditioned eyelid reactions; a study of the reactions of the human eyelid to light and sound stimuli." Orange Park 1930–1932.

Theodore Andrew Jackson. b. November 17, 1901, Muncie, IN. PhD (Psychology) Columbia University, 1931. "General factors in transfer of training in the white rat." New Haven 1932–1934.

Carlyle Ferdinand Jacobsen. b. January 17, 1902, Minneapolis, MN. PhD (Psychology) University of Minnesota, 1928. "A study of cerebral function in learning." Orange Park 1930–1931, New Haven 1931–1937.

Marion Myer Jacobsen. PhD University of Minnesota, 1928. "The quantitative determination of scholastic interests among college students." Orange Park 1930–1931.

William Oliver Jenkins. b. November 7, 1920, New York, NY. PhD Yale University, 1942. "Studies in the spread of effect." Orange Park 1942.

Winthrop Niles Kellogg. b. April 13, 1898, Mount Vernon, NY. PhD (Psychology) Columbia University, 1927. "An experimental comparison of psychophysical methods." Orange Park 1930–1932.

Robert Beverley Malmo. b. October 24, 1912, Panama Canal Zone. PhD (Psychology) Yale University, 1940. "Effects of removal of the visual cortex on brightness discrimination and spectral brightness distribution in the rhesus monkey." New Haven 1937–1941.

Thomas Logan McCulloch. b. August 21, 1905, Centerville, IN. PhD (Psychology) Duke University, 1933. "A study of the cognitive abilities of the white rat with special reference to Spearman's theory of two factors." Orange Park 1935–1937, New Haven 1935–1937, 1939.

Henry Wieghorst Nissen. b. February 5, 1901, Chicago, IL. PhD (Psychology) Columbia University, 1929. "The effect of gonadectomy, vasotomy, and injections of placental and orchic extracts on the sex behavior of the white rat." Orange Park 1939–1958, New Haven 1932–1939.

Vincent Nowlis. b. September 13, 1913, Wilkes-Barre, PA. PhD Yale University, 1939. "Companionship Preferences and Related Behavior in Young Chimpanzees." Orange Park 1940–1942, New Haven 1936–1940.

continued

Table 4.1. *Continued*

Austin Herbert Riesen. b. July 1, 1913, Newton, KS. PhD (Psychology) Yale University, 1939. "Delayed reward and delayed response in chimpanzees." Orange Park 1940–1943, 1946–1949, New Haven 1936–1940.

Kenneth Wartenbee Spence. b. May 6, 1907, Chicago, IL. PhD (Psychology) Yale University, 1933. "Visual acuity and brightness in the chimpanzee and man." Orange Park 1933–1935, 1936–1937, New Haven 1932–1933, 1936–1938.

Sidney Durward Shirley Spragg. b. June 18, 1909, North Battleford, Saskatchewan, Canada. PhD (Psychology) Yale University, 1935. "Displacement of critical responses in stylus mazes by the chimpanzee." Orange Park 1935–1937, New Haven 1934–1937.

Otto Leif Tinklepaugh. b. August 18, 1894, Kalkraska, MI. PhD (Psychology) University of California, Berkeley, 1927. "The delayed reaction in monkeys." Orange Park 1930–1933, New Haven 1931–1933.

Michael I. Tomilin. PhD none. Orange Park 1933–1935.

John Bascom Wolfe. b. July 8, 1904, Dryden, VA. PhD (Psychology) University of Illinois, 1932. "The effect of delayed reward upon learning in the white rat." New Haven 1932–1934.

Robert Mearns Yerkes. b. May 26, 1876, Breadysville, PA. PhD (Psychology) Harvard University, 1902. "The psychic processes of the frog: a study of the instincts, habits, and reactions of the frog by reaction-time methods." Orange Park 1930–1940, New Haven 1930–1942.

Joseph Geno Yoshioka. b. 1893, Osaka, Japan. PhD University of California, 1926. "Weber's law in the Discrimination of Maze Distance by the White Rat." Orange Park 1930–1934.

William Caldwell Young. b. September 8, 1899, Chicago, IL. PhD (Zoology) University of Chicago, 1927. "The influence of high temperature on the guinea pig testis: Histological changes and effect on reproduction." Orange Park 1939–1943, New Haven 1939–1944.

---

[1]Dates of service are approximate, based on annual reports and biographical sources. Numerous inconsistencies between sources are difficult to resolve. Dates in New Haven refer to association with the YLPB. Some, but not all, of those who worked only in the northern branch are included. Numerous New Haven staff members were associated with W. C. Young's research project that was administratively part of the YLPB. It was common for researchers to spend parts of a year in New Haven and Orange Park.

general difficulty of finding good men: "It is imperative that the resident staff be strengthened. . . . But it is difficult to find men of first-rate mentality and character, and very much more so to discover and interest brilliant and entirely dependable investigators" (Annual Report 1932, 6).

He also criticized the overall lack of productivity of his staff, complaining of "the resistance most relatively inexperienced investigators feel for composition" (Annual Report 1935), noting that the typical output of one publication per person per year was inadequate. The situation improved over the next two years (Annual Report 1935, 1936).

Between April and August 1937, however, six of the nine staff members of student assistant rank and above received job offers; four—Jacobsen, Spence, Bruhn, and Spragg—accepted the offers. Nissen and Crawford declined. This effectively reduced the scientific staff to half-strength: "Almost at a stroke it lost an aggregate of twenty-two years of special training and experience in the use of primates as materials of research" (Annual Report 1937, 3). Yerkes was devastated by the "catastrophe" of the "sudden disintegration of a research group which it had cost several years of strenuous and oft-times very difficult and discouraging labor to build" (4).

Surely, one factor in this exodus was the uncertainty regarding the future of the YLPB. In 1937, Yerkes was 70 percent of the way through the initial ten-year funding period provided by the Rockefeller Foundation. Continuation of funding was uncertain. Yerkes anticipated that some of his staff might have to be dismissed in 1939 if a continuation of funding was not forthcoming and he advised his staff members "not to ignore professional opportunities elsewhere" (Annual Report 1937, 20).

Several annual reports include a story of the latest staff member to leave. To some extent, this was a normal phenomenon, with people spending several years in a facility like the YLPB, and then moving into solid, academic positions. Yerkes seems to have interpreted these departures more as defections, perhaps because he saw his investments in training good staff members disappear as the new positions opened up.

By 1939, "the epidemic of uncertainty and unrest provoked by rumors of the prospective discontinuance of the Laboratories subsided" (Annual Report 1939, 18) and the staff now worked more effectively. Yerkes, however, still sought better men for the staff: "with the beginning of the new era it would seem desirable to strive for a higher level of creativeness in the staff, greater maturity and productivity, and to attempt to achieve reasonable stability and continuity of service" (Annual Report, 1939, 19).

Gender Issues

We can use the term of the time, "scientific men," when dealing with the YLPB under Yerkes. There were no women as graduate students or full-time assistants, although Carlyle Jacobsen's wife was listed as a part-time assistant. The first woman staff scientist at Orange Park appears to have been Elaine Kinder, who arrived in 1943, two years after Yerkes retired as director.

The gender imbalance appears to have been a deliberate policy of Yerkes. The best evidence of this is the story told by Eleanor J. Gibson described in chapter 2.

A few female visitors, however, worked at the YLPB for brief periods of time. Noted developmental psychologist Myrtle McGraw, accompanied briefly by John Dewey, spent a two-week period there early in 1935. She observed the interactions of mothers and infants and speculated about possible evolutionary mechanisms of behavior (McGraw 1941). Apparently, Yerkes regarded a woman visitor as an acceptable part of the YLPB scientific effort, although he believed that she "created a difficult situation and considerable friction" (Gregg 1935).

Abraham Maslow

An interesting case of "one that got away" was A. H. Maslow, later a renowned humanistic psychologist and APA President. As he completed his dissertation on rhesus macaques with Harry Harlow at the University of Wisconsin, Maslow (1932) sought a fellowship to work with Robert Yerkes. There was considerable correspondence back and forth but, in the end, Yerkes appeared unable to recommend Maslow for a fellowship because of confusion in processing the proper paperwork between New Haven and Orange Park (Yerkes 1932d).

Winston (1996) pointed to the prejudicial manner in which psychologists of the time wrote letters of recommendation for Jewish students. There are two examples of this in the Maslow-Yerkes correspondence. Clark Hull (1932) in recommending Maslow to Yerkes, wrote that "he is definitely Jewish, with the characteristic Jewish eagerness. I don't know him well enough to be able to state anything much about how obtrusive the less attractive Jewish traits are in his case." Hulsey Cason (1932), wrote that "although he is a Jew, I can assure you that he does not have *any* of the objectionable characteristics for which the race is famous."

Other Staffing

On Transient Scientists

A cornerstone of Yerkes' vision for the YLPB was that it would be staffed by permanent employees—individuals who had a long-term commitment to the facility. It is also true that these were individuals over whom Yerkes could expect to have greatest control.

Yerkes was generally unwilling to have transient scientific staff members at the laboratory. He gave as his reasons that they tended to be disruptive of the general routine. In a 1935 interview in New Haven with Alan Gregg (1935) of the RF, Yerkes went over the various transients who had worked at the laboratory. Some experiences, such as those with F. G. Benedict, Harold Coolidge, and Edgar Allen had worked well. Others, however, had created problems. Robert Hegner had come to study parasites, interfered with the normal feeding of the animals, and disrupted ongoing work. An unnamed NRC fellow had been assigned to the station and resulted in further time wasted. Solly Zuckerman "got lonely and if he had stayed longer would have been more troublesome." Myrtle McGraw "created a difficult situation and considerable friction." Yerkes went so far as to tell Gregg that he would not be interested in continuing the station in the event the model were switched from his vision to one more like Woods Hole, where transient investigators came to complete specific projects.

The Support Staff

An institution such as the YLPB does not run on the work of the chiefs alone; the work of the support staff is also critical. Secretaries, caretakers, and laborers were also essential to the success of the YLPB.

One key member of the staff was Helen Morford, Yerkes' secretary and personal assistant. Ms. Morford migrated between New Haven and Orange Park with the Yerkes family and played an important role in maintaining order in Yerkes' and the YLPB's activities, as is so often the case with a trusted, but underappreciated secretary. She compiled many of the records and edited some of the publications. Yerkes recognized her importance by upgrading her job title to Assistant to the Director in 1934.

The other key staff member was Yerkes' first hire, William C. Atwater, as the initial superintendent. Atwater began his employment the day that the station opened and continued through July 1946, the only employee to outlast Yerkes himself. In his annual reports, Yerkes frequently praised Atwater, as, for example, when he wrote that "Superintendent Atwater discovers each year new ways of being useful to investigators while maintaining grounds, buildings, and equipment in admirable condition and doing far more than his share in maintaining the animal colony intact and serviceable" (Annual Report, 1935, 30).

In 1931, Atwater had the help of just two laborers at Orange Park. During the later 1930s the staff would include three laborers and caretakers. J. Graham Townsend was added on July 1, 1939; he would eventually succeed Atwater and remain with the facility, in Orange Park and Atlanta, for thirty years. A full-time secretary also was added in 1931. Marcia Hewey, who occupied the position during 1935–45, left her unique mark on the station. Elizabeth Foster, described in Nissen's June 1940 report on the chimpanzee Infant Studies Program as "an intelligent negress," was added to the staff as an infant caretaker in 1939; she remained for four years. In addition, there generally was a laborer/caretaker and a secretary in New Haven.

These workers were important cogs in the wheels of the YLPB, but it is difficult to learn much about them as they left little paper trail regarding their experiences.

Funding and Financing the Laboratories

A critical aspect of maintaining an expensive facility such as a primate station lies in its funding. A summary of the income for the YLPB is presented in Table 4.2. By my calculations, some $812,000 was received during the period of Robert Yerkes' tenure as director from 1930 to 1941. As the laboratories generally worked on a July to June fiscal year, it should be remembered that in most of the discussion in this section it is the fiscal year that is referred to (e.g., "1932" refers to the 1931–1932 fiscal year). All of these funds came from private sources; the era of public funding was yet to come.

Rockefeller Foundation

It will come as no surprise that the Rockefeller Foundation provided the primary basis of support for the YLPB. Over three-quarters of the funding during Yerkes' tenure came from the RF. Indeed, the RF allocation was the sole support for the southern branch during the early years.

Table 4.2.
Summary of Funding ($) for the Yerkes Laboratories of Primate Biology During the 1930s

Year	Rockefeller Fndn.	Yale Univ.	Institute of Human Relations	CRPS	Fels Fndn.	Other Sources	Sum
1930	140,000	0	0	0	0	0	140,000
1931	40,000	0	0	0	0	0	40,000
1932	40,000	15,800	0	0	0	0	55,800
1933	40,000	15,600	0	0	0	0	55,600
1934	40,000	15,000	0	0	0	0	55,000
1935	40,000	10,000	5,000	3,600	0	976	59,576
1936	40,000	5,000	10,000	3,600	0	877	59,477
1937	40,000	4,000	10,000	4,157	0	4,100	62,257
1938	40,000	5,254	10,000	3,800	0	1,017	60,071
1939	75,000	5,000	10,000	3,800	0	454	94,254
1940	50,000	5,000	0	11,800	3,000	676	70,476
1941	44,000	5,000	0	3,800	6,000	930	59,730
Total	629,000	85,654	45,000	34,577	9,000	9,030	812,241

As noted earlier, Yerkes' progressivist beliefs were in harmony with those at the RF. He worked to maintain that association once the station was opened. Yerkes wrote to Raymond Fosdick, soon to become President of the RF, extolling the virtues of a book calling for the "establishment of an international university as an instrument for promotion of intelligent world planning" (Yerkes 1931b). Yerkes (1934a) wrote to President Max Mason regarding the Century of Progress Exhibition and its celebration of mastery of the environment in the hope that the next such celebration would be in honor of the mastery of ourselves. He elaborated his views on how enlightened persons might work toward ever-increasing control over the conditions of life—a common theme in much of Yerkes' writing and one often expressed at the RF.

Of course, Yerkes' primary concern at this time had to be with the practical problems of funding the station. The basic budget, excluding building costs, allowed $40,000 per year for the support of the station. Any unexpended funds reverted to the RF. In the early years, Yerkes ran the station so that there were balances reverting to the RF each year. Although this may seem strange to current grantees, who work to spend every penny of funding lest the funding be reduced, it was a part of Yerkes' idealist ethic. This was greatly appreciated by the RF officials, as the Foundation's income had decreased appreciably during the economic hard times.

Late in Yerkes' tenure, the RF allocated $35,000 for a new Laboratory Building, when the new, tiered, five-year budget was approved. The allocations of 1940 and 1941 reflect the increased levels appropriated in the five-year renewal of the basic grant.

It is worth stressing that these funds were fairly flexible. The allocation was to found and maintain a primate station, not to conduct specific pieces of research. The RF had no interest in micromanaging the facility. This general support differs from the support for targeted, individual projects which was beginning to appear even in the private foundations and would later characterize most federal funding.

Yale

Yale began providing some funding for the northern branch with an allocation of $15,800 for the 1931–32 fiscal year, the third year of Yerkes' calendar. During the early years, Yale provided no support for the southern branch.

For administrative reasons at the New Haven institution, the funding from Yale was sometimes allocated entirely in the name of Yale University and in other years, in part, through the Institute of Human Relations.

This appears to have been solely a matter of bookkeeping convenience without practical implications for the laboratory, though this was not always clear to RF officials.

Knight Foundation

The Committee on Research Funds of the Yale School of Medicine administered some funds, including the Knight Fund for research. A total of $700 was allocated for Carlyle Jacobsen's research in 1935. Four years of support for the program totaled $1,900.

Fels Fund

The Samuel S. Fels Fund had a broad mandate with an emphasis on problems of human development. In 1940, Henry Nissen secured a five-year, $25,000 grant from the Fund for his chimpanzee Infant Studies Program. Funding would actually continue for seven years. During Yerkes' tenure, $3,000 was allocated in 1940 and $6,000 in 1941.

The three sets of problems in this large study entailed developmental study, study of genetic factors, and experimental behavior analysis. The project began with the first study, which was basically a normative, longitudinal analysis of behavioral, physiological, and morphological development in chimpanzees.

Macy Foundation

Yerkes got some grant support from the Josiah Macy, Jr. Foundation, an organization with an affinity for psychobiological research. He got $3,000 to support research with S. D. Shirley Spragg on morphine addiction. The networking connections were close, as Lawrence Frank was the Executive Secretary. Yale's Bayne-Jones would become the Director in 1938.

Committee for Research in Problems of Sex

It is not surprising that Yerkes was able to benefit from allocations given by the CRPS, which he had chaired since its inception. Further, there was a substantial overlap of membership between those who served on the YLPB boards, or otherwise affected its development, and the CRPS members, such as Edwin G. Conklin, George Corner, Karl Lashley, Frank Lillie, C. N. Long, and Clark Wissler.

The first allocations to show up on his annual financial statements were in 1935, when there was a grant of $2,400 to Yerkes and Edgar Allen to support the salary of James Elder, and one of $1,200 for secretarial assistance in Yerkes' work as CRPS chair. The secretarial funding would be an annual allocation for some years, through 1940. The progress reports filed with the CRPS show not only the work on hormones, ovulation, and the period of receptivity in chimpanzees, but work in monkeys on such topics as the development of the mammary glands and menstruation, and on human menstrual cycles (e.g., Allen and Yerkes 1937).

During the last two years of Yerkes' tenure as director, William C. Young received CRPS funding of nearly $12,000 for a project on "Structural and Behavioral Studies of the Reproductive Cycle in the Guinea Pig." These sums seem trivial by today's standards in the era of federal grants. It must be remembered, however, that this was in an earlier time. Not only did the dollar go further, but there were no sources of truly large grants. The CRPS was an important source of funding for the support of research on sexual behavior during its period of existence.

Several allocations during the early years of the station from the CRPS do not show up on Yerkes' financial statements. It seems as though he reported primarily the Rockefeller and Yale funding in the early years, and made his reports more comprehensive in later years. During 1931–32, Yerkes secured $2,000 for a study of the "Social life and relation of chimpanzee and other primates, with special reference to sexual behavior, the reproductive cycle, and familial relations," the next year $2,500 for a study of the "Social life and reproductive cycle of the Howler monkey," and in 1933–34, $2,000 for a study of the "Social life and reproductive cycle in primates" (Aberle and Corner 1953). Some or all of this support was for the field research of C. R. Carpenter under Yerkes' supervision.

Abreu Fund

Madame Rosalia Abreu, who had been so important in getting Robert Yerkes started working with chimpanzees, died in 1930. In 1937, her son, Pierre, established the Rosalia Abreu Memorial Fund at the YLPB with an endowment of $10,000 to support the publication of research results (Minutes of the Scientific Advisory Board Meeting 1937). Between 1937 and 1941 the Fund would yield an annual income that was helpful, but always less than $1,000.

Rotating Publication Fund

In 1932 Yerkes established a Rotating Publications Fund to facilitate the publication of reports from the laboratories. As of June 30, 1935, the fund amounted to $1,128.41. The income from the sale of publications credited to the fund for 1935 was $276.48.

In April 1937, nine 16mm silent films were made available for distribution through Instructional Films, Inc. The income from these films was added to the revolving fund (Annual Report 1937).

Expenditures

The biggest expense during the early years was in salaries and wages. Thus, for example, in 1933 some 64 percent of the funds were spent in this way. The 1934 list of expenditures, shown in Table 4.3, was representative.

By 1941, the total allocation to the YLPB had risen to nearly $60,000. The general pattern of expenditures was similar to those of earlier years. Of the unrestricted funds, over 80 percent of the $49,000 available went for salaries and wages. Insurance cost $75, and over $800 was allocated for travel and transportation.

Perspective

Yerkes was fortunate to have a single benefactor willing to provide the long-term support granted by the Rockefeller Foundation. However, as

Table 4.3.
Expenditures During 1934 as Representative of the Yerkes Years

Salaries	$24,612.99
Wages	1,560.00
Advisory Board	286.60
Maintenance and improvement	7,393.77
Maintenance of animals	1,763.50
Apparatus and equipment	1,292.93
Travel and transportation	1,422.99
Publication	300.00
Total	38,632.78
Refund: Rockefeller Foundation	1,367.22
Total Allocation	40,000.00

his tenure as director progressed, it became necessary for him to try to find other sources of support. Surely, this effort detracted from his scientific research. Nevertheless, he was successful in maintaining the facility at an adequate level, though not the level he sought.

Regulations

Perhaps reflecting his military experience, Yerkes was a director who tried to run his operation with a strict militaristic hierarchical structure with clear lines of authority from the top down. It was an approach that did not always serve him well.

In the year of its founding, Yerkes developed a *Code of Regulations* to structure the activities at the laboratory (Yerkes 1937a). Significantly, the code begins with a statement on the importance of cooperation; it was unaltered from 1930 to 1937: "Although individual initiative, independence, and originality are primary conditions of success, the nature of the program of research of these Laboratories is such that teamwork and an exceptional degree of cooperativeness, subordination of personal interests, and consistent efforts to contribute to the furtherance of the general program of research, and especially to the necessary routine work of the establishment, are absolutely indispensable" (1937a, 3). Yerkes clearly valued cooperativeness above creativity.

All photographic and other general records would be the property of the laboratories and would be filed in the office of the director. All manuscripts were to be approved by the director. Publicity would be controlled by the university, with all matters of print, magazines, and publicity funneled through the director.

The director would keep a Director's Log of all important events that make up a laboratory history. Each chimpanzee would have a daybook that would include such information as the animal's habitat, environment, health, diet and feeding, examinations and measurements, work, social behavior and relations, general behavior, emotional expressions, and reproductive behavior. Although the daybook would be the key record of each animal, other forms would be added. These included a Summary Individual Record, Dental Record, Body Measurement Record, Physiological Record, and Behavioral Record. Sample forms for each of the various informational items were provided with the regulations. A summary of the Observational Routine and the various reports required on a daily, monthly, or quarterly basis were spelled out in detail.

The imposition of strict rules and procedures was important as it enabled laboratory personnel to develop a bank of data on the life histories of captive chimpanzees that was unrivaled and useful in many contexts. Yet when Yerkes' management style spilled over into other nonresearch interactions with the scientific staff, problems developed. For Yerkes, "the motto of our establishment might well be, 'Each for all and all for each'" (Annual Report 1937, Appendix A, 4).

The Physical Plant: Buildings, Grounds, and Equipment

The Orange Park station initially was built on a small part of a 188-acre site. The station itself was enclosed within a fenced area that occupied less than an acre. The primary buildings at the time of the opening of the station were a Laboratory, a Service Building, and a Quarters Building.

Numerous modifications would be made over the years. Yerkes loved planning and building new facilities. Elder recalled, "it was my impression that for Doctor Yerkes games and other recreational activities were lively interludes in which he could participate while thinking about the design of a new addition to the laboratories" (1977, 37).

The laboratory facilities for the northern division, in the fifth- and sixth-floor "penthouse" of the new Yale School of Medicine on Cedar Street, were extensive (see Yerkes 1932a). There were large research rooms with individual cages for monkeys, a seminar room, good offices, and a kitchen. Large indoor living rooms and an outdoor cage were located on the sixth floor.

Plantings

The landscaping of the station was a special passion of botanist Ada Watterson Yerkes. Mrs. Yerkes described her work on plantings, and her daughter appended it to some copies of *Yale in Florida* (A. W. Yerkes 1939) when it was bound in 1966. The material includes detailed maps of where the various species of trees and shrubs should be placed. During the 1930s, Ada filed formal annual reports on the condition of the plantings, many addressed formally to Dr. Robert M. Yerkes.

Assorted Additions and Changes

Each year saw Yerkes contemplating or making modifications to his play toy. Almost all of Yerkes' annual reports included news of minor modifications. For example, in 1932, salt containers were added to each cage; in 1938 electricity was installed in the Quarters Building, allowing experimenters to use electrically controlled devices.

In 1932 a tree-rich area of 60 × 70 feet was fenced in on an experimental basis in an effort to provide a seminatural area for the chimpanzees. Some of the adult animals were able to scale the fence and so the effort was terminated.

Beginning in 1931, Yerkes realized that housing facilities were needed for investigators, as there was relatively little housing in the immediate area. He drafted plans for three small residences for investigators and frequently raised this issue in the annual reports to the RF officials, and at Yale. However, these plans never came to fruition: with the exception of the superintendent's house, all investigators were on their own for housing.

Yerkes first mentioned the need for a building for use as a physiological laboratory in his 1935 Annual Report. In 1939, renewed RF funds permitted the construction of the new Laboratory Building; the original Laboratory became known as the Administration Building. The new Laboratory was west of the Administration Building and connected to it by another covered walkway. It contained experiment rooms, offices, a surgical suite, an autopsy-dissecting room, a chemical laboratory, chemical storage room, and a suite planned for anatomical-physiological studies. The original laboratory building was modified so that it then contained administrative offices, a library and reading room, photographic dark room, X-ray room, and a superintendent's office on the first floor, in addition to a soundproof room, observation rooms, and a nursery suite for use in Nissen's Infant Studies Program.

Equipment

The laboratory facility contained assorted basic equipment at its founding but new equipment was added as the program developed throughout Yerkes' tenure. Much behavioral testing equipment would be developed in both New Haven and Orange Park during the 1930s. In 1933, for example, both a token-reward system, a "chimpomat," and a strength-measuring device were added to the testing arsenal in New Haven (Annual Report 1933). An apparatus to measure various dimensions of color vision, designed by Walter Grether, was later added in 1939.

With the renewal of funding from the RF, new equipment was added to the station in 1940, including the outfitting of surgical, X-ray, and anatomical suites, the purchase of some measuring devices, and construction

Photograph of the New Laboratory Building of the Yale Laboratories of Primate Biology. Courtesy of the American Philosophical Society, Philadelphia.

of special cribs, cages, etc. for the nursery. And, in the next year, further additions included sound recording apparatus, an ultraviolet generator, fluoroscopic equipment, a delineascope (used for projecting opaque printed matter, photos, slides, or gels onto a screen), and additional machinery for the nursery suite.

Animals

All chimpanzees in the colony had both names and numbers: all males were given names ending in consonants; females' names ended in vowels. Males carried odd numbers; females even. Each annual report included a summary of additions and losses to the colony.

The initial chimpanzee population in Orange Park consisted of the four animals that Robert Yerkes had purchased in 1925 plus ten brought from French Guinea by Henry Nissen in 1930. By July of 1931, fourteen additional individuals were added. Losses included two by death, two by sale to the Yale Department of Physiology, three moved to the northern division, and one, Jim, on loan to the Philadelphia Zoological Garden.

Upon the death of Madame Abreu soon after the opening of the YLPB, her son allowed Yerkes to pick out ten chimpanzees from her colony in Havana. Yerkes estimated that this advanced the progress of the station by two years. Mr. Abreu provided two additional females in 1932, and an important female, Lita, of known age and parentage, in 1935. The other big event was the birth of Alpha, the first station birth, on September 11, 1930.

There was a total of thirty-six births, equally divided between males and females, during the ten-year period of operation of the station through June 30, 1940 (Annual Report 1940). In a typical year, there would be about four births in the colony. With the initiation of the Infant Studies Program, there was a spurt in the birth rate, with nine births in 1939-40 and five the next year. Infants were generally separated from their mothers as soon after birth as practicable, in order to include them in the Infant Studies Project.

The first pair of chimpanzee twins on record, Tom and Helene, was born to Mona and Pan in Orange Park on June 26, 1933. This generated much interest and some publicity (e.g., *Florida Times-Union* 1934).

With his history in the eugenics movement and interest in genetic influences on behavior, it was natural that Yerkes would plan selective breeding studies (Annual Report 1935, 26). However, there is no evidence that his plan was ever put into practice on an substantial scale.

The health of the animals fluctuated somewhat from year to year. Although primate veterinarians were not available in either New Haven or Jacksonville, practicing physicians sometimes helped the staff with chimpanzee medical problems. Otherwise, the investigators had to learn what they could and do their best with it. Advice from pediatricians and the personnel of the Florida State Board of Health also helped (Riesen 1977). In 1934, Yerkes reported the first three losses to pneumonia. An epidemic of bacillary dysentery, resulting in one death, struck the northern division in September 1935. A respiratory infection spread through the colony of the southern division in January 1936. By 1937, Yerkes reported that the "menace of dysentery and pneumonia is great," but that they had been successful in avoiding any cases of tuberculosis in the colony (Annual Report 1937, 15). There was an outbreak of a virulent respiratory disease in April 1940.

The facility for which Yerkes had worked so long to establish was up and running, yet Yerkes' tenure as director would result in a mix of accomplishment and disappointment.

5
Research During the Yerkes Years

THE PURPOSE OF THE YLPB, AS CONCEIVED BY ROBERT YERKES, WAS to conduct a comprehensive research program on the great apes. Although the output of the YLPB during the Yerkes years was sometimes regarded as less than one might have hoped at its founding, it was still substantial. The reader must remember how little was known about chimpanzee biology and behavior at the time and should realize the comprehensive nature of Robert Yerkes' vision.

The research accomplished will be summarized in this chapter. A nearly complete list of publications dating from the founding of the YLPB until the present has been prepared by the staff at the Yerkes Regional Primate Research Center in Atlanta and is available to the reader. Further, almost all of the research discussed is available in print. Therefore, I shall merely discuss some highlights representative of the work done during the years of Robert Yerkes' directorship.

Although the research was less comprehensive than desired by many of Yerkes' evaluators, it still covered a wide range of behavior and other characteristics of chimpanzees. There appeared to be no single, comprehensive plan. Rather, the idea was to build basic knowledge in several areas of psychobiology. The relationship to Yerkes' social engineering goals is rarely obvious.

BASIC HUSBANDRY, ANATOMY, AND PHYSIOLOGY

Because so little was known about the basic biology of chimpanzees, staff members had to study the fundamentals of maintenance, anatomy, and physiology. This often required considerable retraining as they had been educated to work primarily with behavior. It may have been the supreme accomplishment of the YLPB under Yerkes. In Part 3 of *Chimpanzees: A Laboratory Colony,* Yerkes (1943) provided a comprehensive summary of the care and feeding of captive chimpanzees that was state-of-the-art for the time.

Elder (1937a) developed methods of anesthetizing chimpanzees by training the chimpanzees to cooperate for injections by rewarding them with small quantities of fruit. Other reports contained information on such topics as the chimpanzees' weight norms (Grether and Yerkes 1940) and pathologies (e.g., Elder 1939b). A major coup for the time was the recovery of a perfect chimpanzee embryo approximately ten-and-a-half days old (Elder, Hartman, and Heuser 1938).

Species-Characteristic Behavior

Many of the studies at the YLPB were of behavioral patterns characteristic of the species, many of which appeared to reflect at least some genetic influence. Finch (1941a) found twenty-five of thirty chimpanzees displayed marked handedness. In contrast to data in humans, right- and left-handedness were equally prevalent. Yerkes (1943), who had resisted early attempts to train him in right-handedness, attributed the prevalent right-handedness in humans to cultural influences and believed that "the free natural development observable in the chimpanzee is preferable to the custom-imposed handedness development of civilized man" (115).

Although Yerkes believed timidity, shyness, caution, boldness, wildness, and savageness all to reflect genetic influence, he noted that of the lone pair of twins born in the laboratory, obviously fraternal twins, Tom was bold and Helene was relatively timid.

Field Studies

Yerkes' modest program of field studies may be justifiably regarded as the beginning of modern field primatology. The program was initiated by Robert Yerkes although the actual research was conducted by Henry Nissen, Harold Bingham, and, especially, C. Ray Carpenter.

Nissen was chosen as the scientist to fulfill the cooperative agreement with the Kindia Laboratories. He conducted a two-and-a-month study and, in addition, when he returned in June, brought back the sixteen chimpanzees described earlier (Nissen 1931). Nissen believed that chimpanzees lived in such a favorable habitat, with food, water, and other resources readily available and enemies not often troublesome, that much of their energy is devoted to emotional expression and the more intellectual and cognitive capacities of chimpanzees seemed minimally taxed in securing resources. In this respect, Nissen may have underestimated what has since been revealed.

Bingham traveled to the then Belgian Congo in 1929 (Bingham 1932). Bingham and his party spent most of their time tracking gorillas, with relatively few direct observations. He provided information on the habitat of the animals, their general movements, and especially on the characteristics of their sleeping nests. However, he could not provide much direct information about their actual behavior. Of Bingham's study Allee (1933: 320) wrote that "at last such field studies have been put on a sound basis which should result in the hunting of information rather than specimens."

The man regarded as the "father" of modern primate field research was C. Ray Carpenter (see DeVore 1965); one might argue that Yerkes was the grandfather. Carpenter (1934) studied howler monkeys on Barro Colorado Island in the Panama Canal Zone. He was able to observe these animals more consistently than Nissen or Bingham and found that the howlers live in small groups of about seventeen animals. Carpenter concluded that the howling, for which they are so renowned, functions as a threat. He described the ecology, feeding, sexual behavior, vocalizations, and other characteristics in his very thorough study. Carpenter would remain on the staff for just three years. However, later he would develop a career of sustained contributions to primate field work (e.g., C. R. Carpenter 1940). An interesting perspective on Carpenter's work appears in the writings of Donna Haraway (e.g., 1983, 1989).

Sensory-Perceptual Processes

If we are to understand the behavior of animals of any species it is critical to understand their sensory and perceptual systems. In general, the YLPB staff found chimpanzee sensory function to be roughly comparable to that of humans.

Elder (1934) reported that the thresholds for hearing sounds were generally lower in chimpanzees than in humans, but that they were within the limits of human variation. Spence (1934) found visual acuity in chimpanzees to be roughly the same as that in humans and superior to species such as chickens, pigeons, white rats, honey bees, and fruit flies, though similar to that of cebus monkeys. Walter Grether (e.g., 1939) found color vision in monkeys and chimpanzees to be comparable to that of humans. McCulloch (1941) measured the just noticeable differences in chimpanzees in discriminating lifted weights. As with the other modalities, the sensory capacities were similar to those of humans.

Development

Developmental research was emphasized throughout the life of the YLPB. This work can be divided into studies aimed at defining norms in development and those designed to analyze the effects of specific manipulations.

Normative Studies

A series of normative studies of the growth and development of groups of young chimpanzees was completed during Yerkes' tenure (e.g., Spence and Yerkes 1937). Other developmental studies used what was essentially a "baby-book" approach, in which detailed records of a single animal were reported. One was of the first infant born in Orange Park, Alpha, born on September 11, 1930, who spent her nights for eight months at the home of the Jacobsens (Jacobsen, Jacobsen, and Yoshioka 1932). They provided detailed data on the physical growth and development of Alpha, while Tomilin and Yerkes (1935) focused on the development of the fraternal twins, Helene and Tom, born in 1933. A progress report appeared in the *New York Times* (*New York Times,* 1934).

The focus of Yerkes and Tomilin (1935) was on mother-infant relations and the development of independence. Crawford (1938) focused on the individual differences among the young and developed a rating scale with which to capture them.

The most substantial developmental project dealt with the norms of behavioral development. Beginning in 1939 with support from the Samuel S. Fels Fund, Nissen and his colleagues, especially Austin Riesen, undertook an ambitious Infant Studies Program designed to study the physical and psychobiological growth and development of parasite-free chimpanzees beginning at birth. This was a major, long-term project that provided baselines for many other studies. Data ranged from daily readings of body weight through fifty monthly anthropometric measurements, pulse rates, behavioral tests using a modified Gesell schedule, and various other behavioral records (Nissen 1942).

Experimental Manipulations During Development

Psychologist Harry F. Harlow would later become famous for his studies of "contact comfort," the drive of young rhesus monkeys to cling to soft

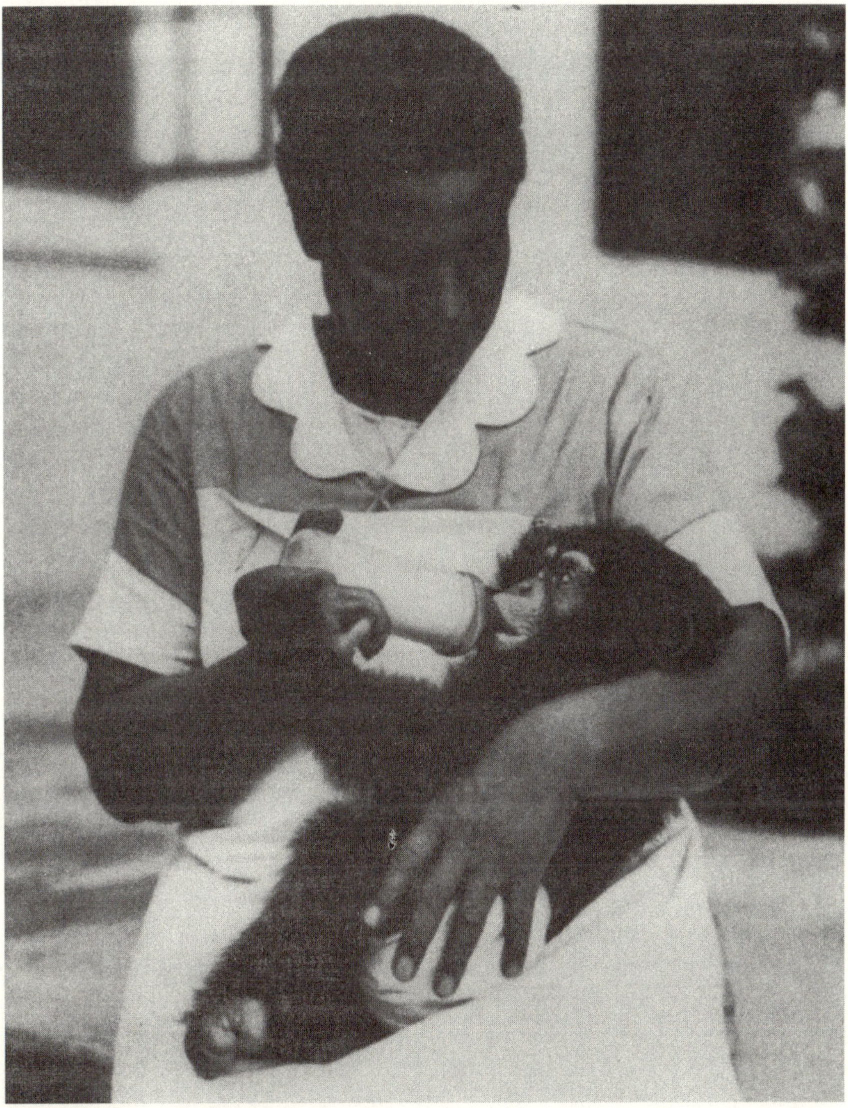

The chief nurse of the nursery program, Elizabeth Foster. From R. M. Yerkes: *Chimpanzees: A Laboratory Colony.* New Haven, CT: Yale University Press. By permission from Yale University Press.

objects, especially in times of stress (e.g., Harlow 1958). Similar observations, though with a different theory, were made by McCulloch (1939) in Orange Park. As later described by Harlow, the apparent need to cling is enhanced in stressful situations. McCulloch found that the opportunity to cling to toweling of the experimenter could serve as a reinforcer in several learning tasks. He developed a theory according to which clasping serves as a reinforcer by inhibiting avoidance, not by removing stimuli.

Home-Reared Chimpanzees

Among the best known and best publicized (e.g., *Time* 1933) studies from the Yerkes years in Orange Park was the study of Kellogg and Kellogg (1933), in which they reared a chimpanzee and their son together in their home for nine months (Benjamin and Bruce 1982; Candland 1993). As with several of the studies discussed earlier, the issue was the nature-nurture problem (i.e., how much of behavior was due to genes and how much to environment). Chimpanzee Gua was born in the Abreu colony on November 15, 1930; the Kelloggs received her at the age of seven-and-a-half months. The Kellogg's son Donald was born on August 31, 1930, and thus was about two-and-a-half months older than Gua. The two "children" were raised under conditions as similar to each other as possible. Donald and Gua were tested daily for nine months with measures of body size, blood pressure, depth perception, reflexes, locomotion, strength, dexterity, play, language comprehension, and a variety of other characteristics. In reading the book written by the Kelloggs, one is struck by their attempt at complete, impersonal objectivity, sometimes seemingly carried to extremes.

The Kelloggs found similarities, as well as differences, of course, between the ape and the child. The major disappointment, of course, was Gua's total inability to begin to learn even the rudiments of human speech. Donald's language development was somewhat retarded during the period of the experiment. The experiment was ended on March 28, 1932 and Gua died on December 21, 1933 with fever and respiratory symptoms. Donald died tragically by suicide in 1973, just months after the deaths of his parents.

Less well known is a similar study conducted by Glen Finch and his family (Annual Report 1938, 26). Fin, as the animal was named, was reared in the presence of the Finch human children. The study was discontinued in February 1940. There are numerous references in correspondence and elsewhere to a manuscript based on this study, but I have found none extant.

Studies in Neuroscience

The YLPB studies from this period with the most impact on society were those of Carlyle Jacobsen and his associates conducted primarily in New Haven in both the northern branch of the YLPB and John Fulton's physiology laboratory (Crawford et al. 1948; Valenstein 1973). The investigators studied the functions of the various portions of the cerebral cortex in monkeys and chimpanzees. The results with the greatest impact concerned ablations of the frontal lobes (e.g., Jacobsen and Nissen 1937; Jacobsen, Wolfe, and Jackson 1935). Jacobsen et al. (1935) concluded that "those activities which demand integration over a period of time can no longer be effectively carried out" (14).

In another important result, the affective aspects of behavior also were altered. In cases where an experimental neurosis had been established by training an animal on a problem too difficult for it to master, after bilateral extirpation of the frontal areas, the animal was calmed and showed none of the emotional responses to failure that had characterized its preoperative activity. The authors pointed to the considerable similarities between the effects of such lesions in humans and in nonhuman primates.

The story has two interesting sequelae. In one, the conclusion that the deficits were due to problems with short-term memory were challenged by several authors, including Karl Lashley, who thought that the performance deficits reflected a difficulty in maintaining a constant set or attitude in the experimental situation rather than a problem of memory (Finger 1994). Further research revealed that the deficits could be minimized with certain alterations of the testing situation, thus supporting Lashley's contention. Crawford et al. (1948) concluded that they had been wrong in attributing the deficits to a loss in memory ability.

Even more interesting is the effect the research had on surgical procedures on humans. In the audience when Jacobsen presented the results of this research at the International Neurological Congress in 1935 was Egas Moniz, a professor of neurology in Lisbon, Portugal. Moniz was interested in the personality changes in chimpanzee Becky, whose symptoms of experimental neurosis were alleviated. Moniz asked whether such ablations might alleviate anxiety and delusional states in humans, a question that "seemed to alarm both Jacobsen and Fulton" (Finger 1994, 291). Moniz found inspiration in the research and, with Pedro Almeida Lima, initiated a program of prefrontal lobotomies in humans. They operated on twenty human patients; Moniz received the Nobel Prize for his work. In the United States, Walter Freeman and James W. Watts picked up the procedure. The number of lobotomies performed increased tenfold between 1946 and 1949, the high-water mark, with Freeman and his associates alone performing approximately 3,500 operations. Although the operations made many patients easier to manage, they produced "armies of zombies" (294). Understandably, the lobotomies became quite controversial. With the appearance of improved tranquilizer drugs in the 1950s, pharmacological intervention became the method of choice and the use of radical psychosurgery declined.

Addiction

Although there was much information on the physiological effects of opiates, Spragg (1940) set out to study the desire for the drugs that is so characteristic of addiction. Chronic morphine addictions were induced in four chimpanzees. Qualitative descriptions of the behavior of the animals about to receive injections revealed a general eagerness to get to the injection room and cooperate with the procedure. They would lead the experimenter to the room and hand him the syringe. Quantitative studies followed. Withdrawal symptoms similar to those in humans were observed. Spragg found various physiological effects, particularly on sexual function, as well.

It had been widely thought at the time that only humans could become addicted to morphine because psychological and social processes were necessary for its development (Laties 1986). Spragg not only demonstrated addiction in a nonhuman species but concluded that "morphine addiction is fundamentally a physiologic phenomenon, developed according to principles of association" (128). He argued against a prevalent theory of the time regarding the importance of societal factors in addiction, as they seemed completely lacking with chimpanzees. "The obvious conclusion from this work was that the chimpanzee resembled man in its reaction to morphine and could serve as a model for man" (36; see also Spragg 1993).

Social Behavior

The topic of social behavior was a favorite of Robert Yerkes. For example, for the Golden Jubilee volume of the *American Journal of Psychology,* he wrote "Primate Cooperation and Intelligence" (Yerkes 1937e), which sounded many of his favorite themes. Yerkes stressed the central place of social behavior in chimpanzee and human life and the importance of the study of social behavior for understanding human evolution.

In what he termed a "naturalistic study of a captive chimpanzee family," Yerkes (1936b, 369) observed the behavior of the male Pan, the female Nana, and their son Don. He described the sexual behavior, parturition, and social behavior in the group. Yerkes concluded that this core family was representative of chimpanzee families in nature, ignoring the fact that he had set up this "family" in the image of the dominant western human families of his time through his manipulation of their caging arrangements (see Haraway 2000). Later field studies suggested the presence of very different mating systems in chimpanzees.

Nissen and Crawford (1936) studied the conditions under which chimpanzees shared food or tokens. Animals would beg and even tease each other, appearing to offer food and then withdrawing it. The research on sharing was one of the stories from the laboratories picked up in the media (e.g., "Yale Chimps, No Chumps, Champs At Begging, Generous Too," *New Haven Register,* Scrapbook).

Perhaps the best known experimental research on chimpanzee social behavior was that of Meredith Crawford (1937, 1941) on cooperative problem solving. The basic paradigm entailed training individual chimpanzees to solve a problem, then altering that problem so that two animals had to cooperate to achieve a solution. In one problem, each chimpanzee was trained to get an incentive by using a rope to pull a box to its cage. The box was then made too heavy for one individual to pull it in. The animals were generally effective in this endeavor. Advanced stages developed spontaneously when one animal would first watch the other, so that it could pull in synchrony with the first and finally when a member of the pair would not only coordinate its efforts with the other animal, but would also solicit cooperation.

Sexual Behavior

I will discuss research related to sexual physiology and behavior, with an emphasis on the nature of estrus and dominance reversals, in special detail because they provide an interesting case study of how research was conducted by Robert Yerkes during this period (see also Dewsbury 1998b).

Estrus in Chimpanzees

The first set of experiments addressed a straightforward question: Do chimpanzees have cycles of receptivity, or estrus, or are females essentially continuously receptive?

In New Haven, Tinklepaugh (1933b) followed one female for eighteen months, concluding that "the copulation period began within 1 to 4 days after appearance of swelling and ended at a similar time prior to its subsidence" (543). That conclusion of a limited period of receptivity would be maintained throughout later work in Orange Park.

Research at the Orange Park laboratories showed that chimpanzees have a menstrual cycle averaging about five weeks in duration. It is divided into six phases: (1) a menstrual phase accompanied by menstrual bleeding, (2) a postmenstrual phase, (3) a tumescent phase during which swelling of the genital region occurs, (4) a period of maximal genital swelling, (5) a period of detumescence, when the swelling is decreasing, and (6) a premenstrual phase.

Yerkes and Elder (1936a, b) studied the sexual behavior of five male and thirteen female chimpanzees at Orange Park at different phases of the females' menstrual cycles. In a "typical" test with experienced, eager, compatible, familiar, friendly partners, the male would display interest in the female using any of several gestures and postures. The female would then run to the male and crouch in a posture of presentation. Insertion would be followed by a variable number of pelvic thrusts and ejaculation by the male.

At the most general level, Yerkes and Elder concluded that the female chimpanzee displays a distinct estrus, with female desire for sex increasing during the tumescent and maximal swelling phases, peaking near the time of ovulation at maximal swelling, then decreasing rapidly, and remaining low during other phases. This period of receptivity correlates generally with the period of peak fertilizability (Elder 1938).

However, Yerkes and Elder also noted that these simple generalizations do not hold for all animals. As was typical of Yerkes' research during this period, they presented table after table of data from individual tests. After highlighting several important cases, they emphasized "the extreme diversity of results of experimental matings and the impossibility of doing justice to the facts by any simple generalization" (Yerkes and Elder 1936a, 17). In an effort to bring order out of these complexities, Yerkes and Elder delineated a set of six classes of variables that account for many of these individual differences. They emphasized the controlling role of the female, noting: "the male is suitor and servitor, not lord and dictator" (10).

My reading of subsequent studies in this line of research led me to coin a new term for the kind of process illustrated by Yerkes and his colleagues as the research program progressed. I term the progressive loss of qualifications and complexities with successive presentations of research results *secondary simplification* (Dewsbury 1998b). The process of *primary simplification* entails the selection of some, but not all, of the data when drawing conclusions within an individual research article. In the process of *tertiary simplification,* other authors simplify the scientist's results.

Yerkes' successive summaries of the results of this research illustrate the process of secondary simplification admirably.

Armed with the new and improved rating sheets, Yerkes (1939a) expanded upon the earlier study. The results were substantially the same in that there was a general tendency for increased copulation during the period of genital swelling compared to other phases of the cycle. However, with the more refined scale, Yerkes could now report that copulations during periods of minimal swelling were more likely to be initiated by the male than by the female. Ratings of female receptivity, desire of the male, acceptability of the female, and acceptability of the male all varied with the cycle.

Yerkes proposed that "the higher the order of behavioral adaptiveness (general intelligence) and the more dominant the male of the species, the wider the range of copulatory responsiveness in the typical sexual cycle and the greater the tendency of the female to respond accommodatingly to his advances irrespective of her sexual status" (79). Yerkes was now looking more broadly and comparatively in relating the frequency of copulation outside of estrus to both the level of intelligence and the extent of male dominance in a species.

Yerkes (1943) again emphasized both the generalization that chimpanzees display an estrus and the fact that the pattern is not always displayed. There had been a subtle shift in tone by this time, however. Yerkes now emphasized the generalization and pointed to the complicating factors as creating what he now treated as distortion of the true underlying relationship. The results, of course, remain unchanged, but foreground and background had been reversed with the generalization assuming primacy over the dynamic complexity.

By the time of his autobiography the qualifications and complexity had completely disappeared. Yerkes wrote simply that "the genital swelling phases (3 and 4) may cover from one to two weeks of the five-week period, and during the greater part of that time the female ordinarily is sexually receptive" (1950, 327).

Thus, we see the process of secondary simplification, as the initial report was laden with qualifications that got stripped away. By the time of the final retelling, the emphasis was on the generalizations, and the qualifications were virtually forgotten. I believe that this process occurs in many fields of science.

CONTEXTUALIZATION OF THE STUDIES

A remarkable characteristic of these papers is the failure to relate them to issues of the time. Yerkes and Elder (1936a) provided no broad context for their research. Yerkes (1939a) did introduce his work in relation to the wise control of human sexual conduct, consistent with his program of "human engineering," but did nothing to return to the broader questions later in the paper. The studies just seem to exist, removed from significant ideas that were then current.

These features might not be taken as remarkable were there not significant issues of the time to which these data were directly relevant. For Malinowski (1927) and others, human sexual behavior differed fundamentally from that in other species. Whereas that of other species was believed to be instinctive and determined by physiological factors, the human male and female were considered ready to make love at any point in the cycle. Human sociality was seen as emerging from the monogamous pair caring for offspring. This view was contested by Miller (1928, 1931), who argued that among primates, like humans but unlike other mammals, sexual behavior was freed from the limitations of the female's cycle. Miller (1928, 1931) and Zuckerman (1932) believed that the opportunity for sex throughout the cycle provided the glue that accounts for the emergence of human sociality. What is puzzling is the failure of Robert Yerkes to address the momentous question of the origins of human society that was being debated with data of precisely the sort he was collecting. Perhaps he was just a conservative scientist unwilling to make such leaps. Perhaps he really believed that he had found a practical way to improve human society. Perhaps this rather straitlaced man was uncomfortable with the matters. Whatever the reason, Yerkes remained silent on these major issues of the day.

QUESTIONS OF METHOD

It would obviously be inappropriate to criticize research done in the 1930s for failing to live up to standards of the twenty-first century. However, there are some characteristics of the work that should have been apparent in the 1930s. For example, there were no validated criteria to determine the occurrence of ejaculation, and when rating scales were used, no indications of reliability or validity were presented.

Although quantitative analyses were becoming more common in the psychology of the time (Danziger 1990), Yerkes' studies include few inferential statistics. Only later would YLPB scientists employ statistics that would permit more precise conclusions (e.g., Young and Orbison 1944). This enabled Yerkes to use his considerable experience in evaluating results but left him prey to his preconceived interpretations.

CONTROLLING THE SEXUAL INTERACTION

Yerkes and Elder (1936a) viewed sexual interaction between experienced, compatible partners, as a *social* process and concluded that "other things being equal, the sexually mature and experienced female controls the mating pattern and determines copulation" (37–38). The dynamic aspects of the interaction were emphasized.

These perspectives also changed subtly in successive summaries, again reflecting secondary simplification. Yerkes (1939) portrayed a greater balance between male and female. For Yerkes (1943) "even when a female takes the initiative by approaching and presenting sexually to a male, the total picture of mating behavior almost invariably indicates a measure of masculine dominance" (71). It is not clear that this shift of perspective was due to new data.

TERMINOLOGY

Another distinguishing characteristic of Yerkes' research was his terminology. We learn that Wendy's behavior "suggests nymphomania" (Yerkes and Elder 1936a, 21), Pan "refuses to accept a female unless entirely satisfied with her condition," (21), and Mona is "large and courageous" (31). As might be expected, Yerkes' friend John B. Watson took him to task for his anthropomorphic terminology rather early (e.g., Watson 1916b). Yerkes (1916f) defended himself, believing that his terminology was scientifically appropriate and that by writing works with such terminology he might interest a broader audience and, in the process, advance the science. After many years in which authors avoided such terms, recently there has been a return to them (e.g., de Waal 1982; Goodall 1986).

Estrus and Dominance in Chimpanzees

The studies of estrus and dominance have been even more controversial. In essence, they concerned power and sex.

THE ORANGE PARK RESEARCH

Yerkes (1939b) studied four kinds of pairings: mates, mature females, mature and immature females, and immature females. In all combinations, except the immature females, Yerkes found evidence of dominance reversals associated with the period of tumescence in the female. Most

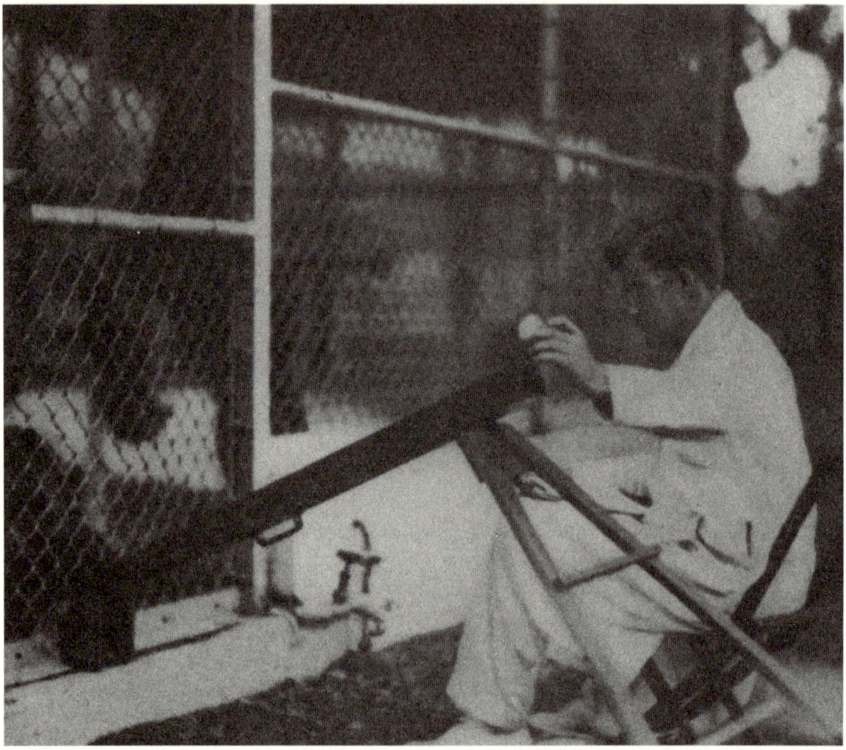

Meredith Crawford conducting a test of dominance with the chute apparatus. From R. M. Yerkes: *Chimpanzees: A Laboratory Colony.* New Haven, CT: Yale University Press. By permission from Yale University Press.

typically, in male-female pairs, the male was dominant (i.e., obtained the food) for most of the tests but the female got the food when in estrus. Yerkes asked the reader to reach these conclusions with a nearly complete absence of data. There is just one table of data from one pair mated for thirteen days. At this point, Yerkes believed that dominance reversals could occur in either direction.

Yerkes (1940) conducted a more systematic study of dominance reversals in male-female pairs composed of four males and fourteen females using a "chute test." Each day, when the chimpanzees were attending, ten pieces of food were introduced into the cage via a wooden trough or chute at intervals of about thirty seconds. The observer simply determined which animal got the food and supplemented these primary data with observations of the interactions between the mates. Yerkes selected nine menstrual cycles that he believed to be appropriate for examination and concluded that, when testing conditions were appropriate, dominance reversals occur. In this study, dominance reversal came to be unidirectional with the typically dominant male "yielding his privileges" when the female was tumescent.

Crawford (1940) followed up this work and was more guarded in drawing conclusions, noting that "the reader has doubtless been more impressed by their variability from pair to pair, than by the evidences for any single relationship which might describe all changes in response priority" (508). Nowlis (1941, 1942) studied pairs of chimpanzees in various combinations of sexes and ages: in chute tests he varied the degree of hunger and sexual status of the two animals. He found the degree of hunger of the subordinate animal to be a factor, but only when the dominant animal was satiated.

QUESTIONS OF METHODS

Again, Yerkes, as opposed to Crawford and Nowlis, relied on his experience and judgment rather than statistical tests.

"YERKES' TRANSFORMATION" AND PRIMARY SIMPLIFICATION

According to Stuart Altmann (personal communication, November 30, 1993) laboratory lore among primatologists had it that a procedure satirically referred to as "Yerkes' transformation" revealed how Yerkes selected data consistent with his hypothesis about this phenomenon of dominance reversals. A study of his treatment of these data provides insight into his approach.

Yerkes (1940) made a total of 8,450 observations on a total of forty-five menstrual cycles. He first eliminated thirteen of these cycles because they include just "fragmentary data for early or late cycle phases" (160) and tests with females known to be pregnant. He then provided raw data for thirty-two cycles. In a later table, he selected "representative data" for each of the six phases of the thirty-two chosen cycles. Yerkes next trimmed the number of cycles he analyzed as prone to seven sets of variables that can distort the true, underlying phenomenon of dominance reversal. Yerkes was now left with just nine cycles of valid data: 8,450 observations had been reduced by more than 80 percent. In his figure 2, Yerkes plotted the data from the selected tests in the thirty-two cycles, the nine chosen cycles, and the positive first trials of each female. The resulting figure is consistent with Yerkes' hypothesis. However, the data in some of the cycles chosen are about as problematic as those eliminated (see Dewsbury 1998b). This provides a nice example of primary simplification.

SECONDARY SIMPLIFICATION

Yerkes' creative graphsmanship continued. In order to eliminate much variability in his graphs, Yerkes presented graphs of how the data *ought* to have looked. These were figures that became associated with his research. It is important to note that these graphs are largely a product of Yerkes' fertile imagination (Yerkes 1939b, 125; 1940, 178; 1943, 73) but they became the archetypical summary of the experiment. The issue at hand is not the veracity of Yerkes' judgments but rather the manner in which Robert Yerkes practiced his science.

LOOKING TOWARD HUMANS

Now, Yerkes interpreted his data in relation to humans. He saw his data as especially useful for "psychopathologists who are intent on appraising, perfecting, and using psychoanalytical method of observation and interpretation" (Yerkes 1939b, 130). He appears to be aiming at Freud! Yerkes suggested that sex is less important in personality development than suggested by the psychoanalysts. In these studies, however, Yerkes found many similarities between humans and chimpanzees, including affective attachments and "such varieties of maladjustment as masturbation, rape, and prostitution" (Yerkes 1939b, 131). One wonders how this supports the argument that sex is overemphasized.

Yerkes finally addressed the writings of both Malinowski and Miller and opposed the view that in nonhuman animals behavioral patterns are common to all members of the species. He agreed that there may be differences between chimpanzees and humans, but asserted that they are differences of degree, not of kind. Again, he tried to occupy the middle ground. Yerkes (1950) saw in these studies the origins of prostitution: "the mature and sexually experienced female obviously trades on her ability to satisfy the urge of the male, and sexual accommodation is, or may be, exchanged for such personal advantages as obtaining food or other desired objects, opportunity for initiative, freedom of action, or leadership" (331). He saw in the granting of privilege "intimations of conscience" (334).

TERMINOLOGY

As with his studies of estrus, we find Yerkes' studies of dominance reversal laced with terms that many would regard as anthropomorphic, such as "courage," "self-assertiveness," "confident," and "venturesome" (Yerkes 1939b, 127).

It is Yerkes' use of the terms "right" and "privilege," however, that became most controversial. For Yerkes (1940), dominance was the ability to gain priority of access to resources; "right" concerns behavior that conforms to social custom; and "privilege" was action that is dependent upon the deference of a mate. Behavior reflecting a right was characterized by positiveness and self-assurance; that of "privilege," by hesitancy and tentativeness. Crawford (1993) recalled that he was uncomfortable with such terms and he and Yerkes agreed to publish their work separately.

FEMINIST CRITIQUES

The most famous feminist critique of this research was that of Ruth Herschberger (1944, 1948), who used the clever device of inventing testimony from the mouth of Josie, a chimpanzee, to dramatize her objections to Yerkes' research. Yerkes referred to the male as "naturally" dominant. Herschberger objected not only that this overstated the consistency with which males actually dominated females, but that it implied that the estrous period was "unnatural." Yerkes (1944) replied that it was the most appropriate word he could find.

Herschberger had some methodological criticisms and was also concerned about the paucity of female scientists at the time and how that affected the conclusions. She viewed the belief that females do not enjoy sex as leading to the argument that the behavior suggested the origins of prostitution. Her most convincing arguments were against Yerkes' terminology, particularly the concepts of right and privilege, and the effect it might have when humans might use his observations to structure their gender relationships.

Several other feminist critiques are relevant to this work but not limited to one set of experiments. Dagg (1984) accused Yerkes (1939b) of acting "unscientifically in reporting his findings" (120). She objected to a passage in which Yerkes suggested that cultural patterns in both chimpanzees and humans led males to develop dominance, leadership, originality, and creativeness, and females to develop subordination, imitativeness, and conventionality. Dagg used Yerkes' own descriptions to show instances in which females displayed traits of the sort Yerkes attributed to males.

Morawski (1988) characterized Yerkes' research program as designed to uncover "the truth of female submissiveness, male domination, social hierarchies, fixed personality differences, chemical addictions, and maternal affection" (81). The latter two characterizations are beyond the scope of the present study. Although there is reason to believe that Yerkes had preconceived notions of male dominance, there is less evidence that he accepted fixed personality differences.

The feminist who has devoted the most attention to Yerkes and the activities in Orange Park has been Donna Haraway (1978; 1989), who focused on the social engineering aspect of Yerkes' work. She highlighted Yerkes' attempts to link sex and power and to naturalize the phenomena that are observed in human society. Haraway presented the research on dominance reversal as a "classic study of the origin of the body politic" (Haraway 1978, 29).

Conclusions

At one level, they are simply experimental studies of chimpanzee social behavior. However, for Yerkes, these studies were pilot studies for the great progressivist project of human engineering. The chimpanzee was an "open book," in which fundamental processes that are masked in humans could be efficiently investigated. These studies could serve as a representation of the relationships between male and female in both chimpanzees and humans and could play a role in shaping attitudes toward important issues of gender relations in humans. Thus, these seemingly simple studies of the periodicity of mating and conflict over small pieces of banana, conducted in the 1930s in an out-of-the-way corner of northeast Florida, had broad implications for the construction of gender-power relationships in human society, and have been interpreted by various authors in that context.

Studies of Learning and Higher Processes

One of the major themes at the YLPB was interest in learning and higher processes. Much of this work was done during a period when many authors have suggested that, under the influence of behaviorism, studies of cognitive processes were nonexistent. However, such research was indeed conducted in the hands of Yerkes and others (Dewsbury 2000).

In the tradition of Thorndike (1911) and behaviorists who followed him, learning was treated as a process of trial and error in which stimuli became associated with responses. Thorndike thought that this single process characterized all learning in all species, except that guided by language in humans. According to an alternative view, other, more complex, processes are involved in some changes in behavior. Such phenomena have often been treated under the rubric of "higher processes," such as concept formation, insight learning, reasoning, and ideation. The central theme is that the accomplishments observed in the animals could not be explained with basic learning processes. Yerkes accepted such interpretations.

Learning

YERKES' TAXONOMY

Yerkes (1943) concentrated on five determinants of learning: attention, motivation, complexity and configuration, temporal relation, and suggestion and imitation. With respect to motivation, Fletcher (1940) found that, within limits, chimpanzees would work harder and longer for a large rather than for a small piece of banana. By "complexity and configuration" Yerkes meant that situational or contextual factors influence performance in learning tasks. As one example, he pointed out that the distance between two stimuli to be discriminated affected the rate of learning (Nissen, Carpenter, and Cowles 1936).

The study by Crawford and Spence (1939) is a classic in the study of imitation. Chimpanzees designated as imitators got to observe chimpanzees designated as demonstrators as the latter performed in a discrimination learning situation between pairs of stimuli. Results were mixed, with just one of eleven animals showing imitation under the most stringent conditions in which there was no reward given to the imitator during demonstration trials.

ABSTRACTION

Yerkes believed that the animals are capable of "abstraction, generalization, and practical judgment" (Yerkes and Yerkes, 1929, 372). Perhaps the most interesting, though not the most complex, aspect of this capability is their general adaptation to testing conditions. Yerkes noted first that as animals are exposed to various testing conditions they get better at performing the tasks. He went further and noted "a transfer of training from a particular experiment to others which are similar in principle, although differing in perceptual values" (372). Yerkes (1943) termed this phenomenon "learning to learn" (130). This is interesting in that the study of learning to learn, or learning sets, would become an important topic as pursued later by Harlow (1949). Yerkes' use of the term antedated Harlow's.

THE WORK OF KENNETH W. SPENCE

Although Kenneth Spence was a doctoral student of Yerkes and spent three years in Orange Park, it was Clark Hull who most influenced his work. The development of learning theory during mid-century was a major enterprise that was more central to the psychology of its time than

Kenneth W. Spence using the Bingham manual multiple choice apparatus. From R. M. Yerkes: *Chimpanzees: A Laboratory Colony*. New Haven, CT: Yale University Press. By permission from Yale University Press.

to that of the present day. Spence was a very influential psychologist within that endeavor. The theory that preoccupied him during much of his life emerged from the Orange Park years.

Spence adopted a more conservative position on matters of learning and cognition than did Yerkes. In Spence's (1936) classic article on the nature of discrimination learning, for example, he argued that some of the phenomena of discrimination learning between two stimuli could be explained with basic reinforcement theory and did not require postulation of more complex processes. The development of Spence's theory was one of the achievements of the YLPB during the Yerkes years that had the greatest impact on the psychology of the time.

Memory, Foresight, and Insight

The concept of ideation was most characteristic of Yerkes' approach. Yerkes (1943) selected four primary areas of research as typifying the appearance of ideation: the string test, the box test, the selective transportation test, and the multiple-choice test. Several more can be added.

THE PATTERNED STRING TEST

In the research of Finch (1941b), eleven problems were used in each of which a chimpanzee had to choose between two or more strings, one of which was attached to a piece of food. The problem was made difficult because the strings crossed in various patterns, so that animals had to use care in determining which one was connected to the food. Seven of Finch's eight chimpanzees solved all eleven problems. Yerkes (1943) interpreted these results as a demonstration that, contrary to the suggestions of some authors, in some problems chimpanzees are superior to monkeys and, like humans, respond in a cognitive-perceptual manner.

THE BOX TEST

In the famous box test, a lure is suspended from the ceiling and the animal must correctly stack one or more boxes directly beneath the lure in order to reach it. (see Köhler 1925). Bingham (1929b) believed that solution entailed novel combinations of behavior that had been practiced in other situations. As evidence of ideation, Bingham listed "(a) abrupt changes; (b) reflective pauses; (c) anticipatory looks; (d) confluent acts; (e) transfers of skill and plan; (f) overnight solutions; [and](g) transformation as revealed in corrective adjustments" (56).

THE SELECTIVE TRANSPORTATION TEST

The Selective Transportation Apparatus (Bingham 1929a) differed from most other tests in that the incentive, such as a banana, was inside of a cage, 6 ft. square and 3 ft. high, and the chimpanzee was on the outside looking in. The incentive was suspended from the ceiling on a rod that ran on a track or channel set in the ceiling of the cage. A knob was fixed to the top of the rod. The animal's task was to move the knob along the track so that the incentive would be moved to one of several small doors where it then could be reached. Bingham was impressed with the versatility of the animals, as they showed abrupt changes, sudden initiation, and correction of errors; he believed ideation to be involved.

THE MULTIPLE-CHOICE TEST

Clearly, Yerkes' favorite method was that of his own devising—the multiple-choice test (Yerkes 1916d, 1934b). Studies were done in a large, outdoor apparatus in which the animal moved through an entrance alley to a chamber onto which opened nine doors, each leading to a large box. On any given trial, some doors were open and some closed. The animal's task was to pass through the correct door and box to obtain food. Within a problem, which included a series of ten trials, which box was correct was determined by a consistent rule, such as "the middle door" or "the second from the right end." Because a different set of doors would be open on each trial within a problem set, the actual door that had to be selected varied from trial to trial (e.g., left-most door among those open). Yerkes regarded the problems as "relational."

Yerkes (1934b) studied four chimpanzees in this apparatus and concluded that the solution resulted from the "sudden discovery of the significant relationship" (103) because of the abrupt changes in error rate from 30 percent or more to zero, the fact that different boxes were open on different trials, and the ability of the subjects "to respond correctly with ease and assurance" to new settings of open doors. He believed that these forms of behavior "obviously presage those expressions of human curiosity and originality we call invention and discovery" (107). Skeptics were not convinced (e.g., Spence 1939).

With Yerkes' encouragement, Spence (1939) took up what was conceptually the same problem, albeit with a new, smaller apparatus. Although performance was even better than in previous studies, Spence remained unconvinced of Yerkes' ideational interpretation.

INSTRUMENTATION (TOOL USE)

Another situation in which Yerkes believed that chimpanzees showed ideational learning was the box-and-pole test of Yerkes and Spragg (1937). A banana was placed out of reach inside a long, narrow box with open ends that is anchored to the ground. The only way for the animal to get the banana was to use a long pole to push it out the opposite end. Yerkes and Spragg particularly noted the performance of one chimpanzee, Mamo, whom, they felt, showed every evidence of having solved the problem suddenly as if with insight.

In work done in New Haven, Jackson (1942) used stick tests similar to those used by Köhler (1925). The chimpanzees had to use a short stick to rake in the next longer stick up to as many as four sticks, only the longest of which was sufficient to rake in the incentive. Jackson contrasted the performance of younger animals, who seemed to act via trial and error, with that of older, experienced animals whose solutions "tended to be of the 'insightful' (sudden) type" (234).

Language and Symbolism

At a still more abstract and cognitive level, Yerkes and his colleagues sought evidence of the use of signs and symbols in chimpanzees as precursors of linguistic capacity. Whereas symbols were regarded as capable of mediating delays between the designation of a correct stimulus and the opportunity to make a response, signs were not. They found evidence of what Yerkes (1943) regarded as real, but rather minimal, use of symbols.

Yerkes was somewhat ahead of his time in believing that, although chimpanzees lacked the vocal apparatus to talk, they might be taught a language in which they would use their fingers, as do the deaf (Yerkes and Learned 1925); this was later found to be true (e.g., Gardner and Gardner 1969).

TOKEN REWARDS

Among the best-known learning studies from the YLPB during this period were those of token rewards; Yerkes (1943) treated these as indicative of sign learning. In the studies conducted by Wolfe (1936) and Cowles (1937), chimpanzees worked for poker chips, as tokens for food, by manipulating a lever in a special apparatus. Later, the animals could exchange the tokens for food by using a "chimpomat" vendor apparatus.

The chimpanzees readily learned the task and discriminated among chips assigned different reward values. When the animals were placed in social situations the tokens elicited competitive behavior similar to that displayed over food. One animal even begged for tokens. Tokens were found to support learning in various tasks. Wolfe concluded that the tokens "came to function as secondary or surrogate rewards" (72).

OTHER STUDIES OF SYMBOLIC PROCESSES

Yerkes and his associates also sought evidence of symbolic processes in studies of delayed response, discrimination learning, and delayed reward. Nissen, Riesen, and Nowlis (1938), for example, reported the results of several related experiments all of which involved variations on a basic discrimination learning problem requiring differentiation between a white and a black panel, one of which led to food. They believed that a symbolic mechanism was necessary for the subjects to solve the problem as they did.

A FAVORITE ANECDOTE

Yerkes (1943) retold one of his favorite anecdotes in support of the view that chimpanzees understood complex relationships. Moos had been ill and, suspecting a problem with the teeth or gums, an experimenter (Nissen) entered the cage to make a dental inspection. Moos cooperated. Failing to find any problem, the experimenter turned to leave the cage. "Moos took hold of his coat, drew him back, and raising his upper lip with one hand pointed with a finger of the other hand to a spot on his upper jaw" (192). There proved to be a swelling at that point and a permanent canine tooth soon erupted, suggesting that this was the cause of the discomfort.

Conclusions Concerning Learning and Cognition

Robert Yerkes concluded that chimpanzees differ qualitatively from both other primates and from humans. Looking "monkeyward," Yerkes (1943) saw a big gap, and believed that "creative imagination, learning processes, and linguistic and emotional expressions are far more varied and highly developed in the chimpanzee." (195) Looking "manward," Yerkes concluded that "the ape is at the beginning of a road on which man has advanced far" (195).

What is most significant about this work is the extent to which it differs conceptually from much that was in vogue at the time. The dominance of behaviorist thinking during this alleged fallow period of comparative cognition was less pervasive than is portrayed in received histories of the time. Robert Yerkes was a very determined, indeed perhaps stubborn, man who would rarely be swayed by the opinions of his peers. The study of animal cognition did not die in the 1930s; it was kept alive in Orange Park and in various other laboratories around the country.

Perspectives

Robert Yerkes (1950) looked ever toward humans. He seemed pleased with the progress made in the laboratories and focused on the implications of the work for understanding human behavior, particularly in four realms: foresight, altruism, conscience, and honesty. What could be more Yerkesian?

Yerkes achieved his dream of a primate laboratory in 1930 and remained as its director for a little over a decade. As I have shown, he encountered many problems. Some of these, in retrospect, seem to have been of his own making. Nevertheless, one must give Yerkes and his staff credit for some significant achievements. First and foremost, they developed procedures for the maintenance, breeding, and study of captive chimpanzees in a subtropical station in the United States. This went well beyond the important accomplishments of Rosalia Abreu. The Orange Park colony was a proving ground for many procedures that would later become standard in the care and maintenance of our closest relatives.

They developed much information on the basic biology, sensory function, reproductive systems, behavioral patterns, growth, physiology, and anatomy of chimpanzees.

In the field studies of Henry Nissen, Harold C. Bingham, and C. R. Carpenter, we have the beginnings of the modern study of primate behavior under field conditions emerging from the Yale Laboratories of the 1930s. A start was made toward a long-term program dealing with the development of behavior. Such concepts as what would later be called "contact comfort" were clearly formulated here. Insight was gained regarding the capacities and limitations of chimpanzees as related to speech. Developments with potential clinical impact were the program of research on the function of the frontal lobes and Spragg's study of morphine addiction.

In the study of learning and higher processes, the Orange Park laboratories were a major center in keeping alive a cognitive approach to learning and the study of complex tasks in an era dominated by research on laboratory rats generally tested in simpler tasks. Taken together, these scientific accomplishments are not inconsequential.

6
People and Life at the Yerkes Laboratories in the 1930s

IN THIS CHAPTER I DISCUSS SOME OF THE PEOPLE WHO INHABITED the YLPB, their lives and recollections, and some of the problems encountered.

TWO PERSONNEL PROBLEMS

Robert Yerkes' tenure as director of the YLPB was plagued and compromised by several characteristics and events. Two personnel problems, those of Harold Bingham and Otto Tinklepaugh, were especially troublesome and had a major impact on the future of the YLPB.

The Case of Harold Clyde Bingham

BINGHAM AND YERKES

The association between Yerkes and Bingham was long-standing. Bingham first worked with Robert Yerkes during his years at Harvard. He worked with Yerkes during World War I, then in the Research Information Service of the National Research Council. He then joined Yerkes at Yale (Bingham, 1969). Bingham accompanied Yerkes to Havana during the summer of 1924. The ties were cemented during summers in Franklin, New Hampshire (A. W. Yerkes 1957). Clearly, Bingham was close to Yerkes.

While in New Haven, Bingham conducted a variety of studies and experiments (e.g., Bingham 1929a,b). His best-known project was in leading an expedition to Africa in an effort to study the behavior of gorillas in their native habitat (Bingham 1932). Bingham's publications were the result of substantial studies that demonstrated Bingham's ability to complete major research projects.

THE DEPARTURE OF BINGHAM

Early in 1930 Yerkes needed someone to fill in and run the station for the rest of the year. Bingham, his trusted associate, was his man; Yerkes wanted Bingham to continue at his then-present rank and salary, but under the auspices of the new Station rather than the New Haven laboratory. Bingham declined the offer (1930a).

Yerkes attributed Bingham's failure to accept the position as due to the fact that he was "in very bad physical and mental condition" (Yerkes 1930c). Yerkes regarded Bingham's failure to accept the appointment as an "embarrassment within comparative psychobiology." He then turned to Carlyle Jacobsen to fill the slot that would have been occupied by Bingham.

For his part, Bingham explained his position clearly (1930b). He had several concerns. First, Bingham supported Yerkes' conception of the anthropoid station as a three-pronged program. Bingham wanted to continue with field research, but sensed a lack of "evidence of academic support to this field work in relation to the general program." Second, he was forty-two years old at this point and wanted more security for his family than would be provided by annual appointments on the new Rockefeller grant at his same salary. Bingham sought "home and income"; he needed long-term security, noting that "the annual upsets over reappointments are no longer worth the risks." Bingham wanted to do research, not become Yerkes' designated administrator in Orange Park (1930b). Bingham pronounced himself in fine physical condition. Yerkes could offer only year-to-year contracts and a year split between New Haven and Orange Park, not field research. A month later, Bingham (1930d) emphasized his admiration for Yerkes' "administrative fairness, personal responsibility, and generous guidance" but repeated that he sought a career in research, not administration. Roswell Angier (1930), chair of the psychology department, was sympathetic to Bingham's case.

President Angell talked with Bingham and wired Yerkes asking him if he would approve a three-to-five-year appointment, with possible salary increases in the later years, and an agreement to work out arrangements whereby Bingham might do field work (Angell, 1930). Bingham declined any possible offer before Yerkes had a chance to respond (1930c).

In Yerkes' March 25 letter to Angell (Yerkes 1930b) may lie the beginnings of Yerkes' problematic reputation for dealing with his staff. As noted, Yerkes developed a reputation for supporting staff members as long as they played by his rules, and dropping them when they began to seek independence. Now, Yerkes did an about-face and belittled Bingham's administrative prowess (1930b). He added that Bingham "has failed to convince me of such qualifications . . . although he is eminently suited for practical work." Yerkes added that Bingham "has not improved in mental ability and as investigator during recent years." Yerkes indicated that he was "extremely disappointed and regretful of Bingham's failure to qualify, and quite as much so for my slowness in admitting the facts."

AFTER YALE

Relations between the Bingham and Yerkes families remained cordial after Bingham's "retirement": One summer Sunday evening in 1932 the Yerkes family was at the Binghams', with the young people telling stories of New England winters (A. W. Yerkes 1957).

A year later, however, Bingham (1933) was still unable to find another institutional appointment, so he wrote to Yerkes telling him that his situation had changed and that he now would like to rejoin him in the laboratories. Yerkes politely rejected the overture (1933a). He did, however, provide a letter of recommendation for a Guggenheim fellowship that would enable Bingham to study gorillas at the San Diego Zoo. In a later letter of recommendation, he would call Bingham "neither brilliant nor disconcertingly imaginative" (Yerkes 1936a) but affirmed Bingham's dependability.

Although Bingham never landed another university position, he had an active career in applied psychology. Many years later, Yerkes and Bingham would reflect upon the events of 1930 in an exchange of letters. Yerkes (1955c) agreed that Bingham probably chose wisely in not going to Orange Park, but he now believed this because he felt that Bingham's family would not have adjusted well to the South. Bingham now explained his actions of 1930 in relation to a "loss of faith in my intellectual alertness," mental illness, and depression (1955a). Bingham reflected that he was pleased with his life and indicated that he had been able to grow in the environment in which he had placed himself. Yerkes (1955d) continued the healing process, explaining the difficult position in which he had been placed. Yerkes wrote that he had thought Bingham had some mental health problems, but had thought it had been "neurasthenia rather than a mental breakdown." Bingham (1955b) appreciated Yerkes' explanation as correcting some misinterpretations that Bingham had harbored for many years.

PERSPECTIVE

An evaluation of these events hinges on the accuracy of Yerkes' evaluation of Bingham. If he indeed lacked the creative spark to be a productive scientist, then perhaps Yerkes was justified in refusing to make the commitment that would bring him to Orange Park. If, on the other hand, as was later alleged, this was the first in a series of cases where Yerkes let go any scientist who began to show independence, then a different interpretation follows. In any event, the stated rationale for Yerkes' behavior changed from time to time.

Clearly, Bingham was highly thought of among his Yale colleagues, including Angell and Angier. We cannot know whether Yerkes was right in not working harder to keep Bingham. What is clear, however, is that, rightly or wrongly, this was the beginning of Yerkes' reputation for dropping those scientists who showed the spark of independence from him.

A Tinklepaugh Tale

Because of the impact of this case, it merits detailed treatment.

OTTO LEIF TINKLEPAUGH

Otto Tinklepaugh was an active researcher; the official publication list from the YLPB includes his name on nineteen publications dealing with such subjects as apparatus, heart rate, self-mutilation, reproductive behavior, physiology and anatomy, diet, parturition, complex learning processes, and even maze learning in turtles.

Yerkes appears to have had considerable confidence in Tinklepaugh's reliability. Tinklepaugh was entrusted with a task of historic import for the YLPB; it was he who was responsible for transporting the initial four chimpanzees from New Haven to Orange Park to open the laboratories. Tinklepaugh then stayed on to help get the Orange Park laboratory running. He then was put in charge of the breeding colony, filling in where Bingham might otherwise have served (Annual Report 1932, 15).

THE DISMISSAL OF TINKLEPAUGH

In May 1932, Yerkes informed Tinklepaugh that his employment not be continued at the station and that he should seek employment elsewhere

(Yerkes 1933e). The matter would flare up into a minor scandal the following year. There were two major issues: that of Tinklepaugh's overall promise as a scientist and the "chimcracker episode," along with other minor ones.

TINKLEPAUGH'S COMPETENCE

Yerkes contended that "at no time during his connection with these laboratories have I felt that Mr. Tinklepaugh was a likely candidate for permanent appointment" (1933c). Yerkes was emphatic that *this* was the reason for Tinklepaugh's termination: "The reason for discontinuance of his services is his failure to measure up to the standard of research ability which has been set by the administration of the laboratories. What has been referred to as the Chimcracker episode [see below], instead of being [the] primary cause for discontinuance of appointment, is an incident, which, however, has significance" (Yerkes 1933e).

In defense of Yerkes, Angier, chairman of the Yale Department of Psychology, confirmed that he had a memorandum in his possession indicating that Tinklepaugh had been aware of his imminent dismissal as of June, 1932 (1933).

In his early annual reports, however, Yerkes had provided favorable comments about Tinklepaugh. He wrote of Tinklepaugh that "during the absence of the Director [he] has conducted the Station most satisfactorily" (Annual Report 1931, 15). Of the improved health of the animals, Yerkes commented, "For this encouraging change in status Dr. Tinklepaugh is chiefly responsible" (16).

Both in 1931 (16) and later (Annual Report 1932, 21), Yerkes worried about Tinklepaugh's lack of background for dealing with the problems of diet and parasites in the chimpanzees, but he gave him extra credit for his successes under such circumstances. Indeed, it was Yerkes who had relied on Tinklepaugh to do work for which he had had little preparation.

By the time that Yerkes prepared his annual report of July 15, 1933, soon after the events detailed here, his approach changed and he repeatedly criticized his former research associate and acting director. Tinklepaugh's involuntary retirement from the laboratory, Yerkes wrote, "implies a practical administrative judgment that a somewhat different type of research equipment and interest and a higher order of originality and ability in research than his are essential if the Station is to justify its existence" (Annual Report 1933, 7).

THE CHIMCRACKER CAPER

A critical aspect of the maintenance of the chimpanzees entailed providing an adequate diet. This had been an ongoing problem with Billy, Dwina, Pan, and Wendy even before the move to Orange Park (Bingham 1926; Cowgill 1927). The laboratory personnel developed a special formula, that was modified by Tinklepaugh, that could be molded into hard-baked cakes called "Chimcrackers." These were baked by a Boston firm.

The "chimcracker episode" is referred to repeatedly in the correspondence among Yerkes, Angell, Fulton, and others. First, Tinklepaugh published an article describing the formulation of chimcrackers as an appropriate principal diet for chimpanzees and monkeys, providing both a recipe and an analysis of the protein, carbohydrate, and fat content of each of the constituents (1931). The latter published data were erroneous, containing incorrect values for the whole milk and molasses components and it was necessary to publish a correction (Tinklepaugh 1933a). Surely, this was an embarrassment to a sensitive man like Robert Yerkes.

The matter first flared up when Henry Nissen and Carlyle Jacobsen noticed the errors in the original article. The basic problem was that no account had been taken of water loss during baking and, as a result, the values shown by Tinklepaugh had been incorrect. In February 1932, the pair corresponded with Tinklepaugh regarding the errors (Jacobsen and Nissen 1932), terming their enclosed recalculations of Tinklepaugh's analysis "misery sheets" and presenting several pages with both Tinklepaugh's and their calculations. Jacobsen (1932) added several points three days later. Tinklepaugh replied briefly acknowledging that his report had been erroneous (1932c). After another prompt and no satisfactory reply from Tinklepaugh, in May Nissen and Jacobsen (1932) wrote to Yerkes concerning the problem, of which they had already made Yerkes aware, and enclosed copies of the correspondence, which Yerkes read for the first time. Yerkes (1932e) recognized a need for new weight determinations and looked to his lieutenants for advice. Somewhat later, they provided revised data and especially recommended that an expert dietician be consulted (Nissen and Jacobsen 1932b). There is no indication that this was done.

In a confidential letter of May 29, Yerkes (1932e) informed Nissen and Jacobsen that he had decided to dismiss Tinklepaugh:

> I told Dr. Tinklepaugh that in my judgment as Director of the Laboratories he had proved his incompetence to an extraordinary degree, and also his un-

> wisdom. That I considered his conduct inexcusable toward you gentlemen and toward me, as well as contrary to the interests of the institution. That I should expect to relieve him of administrative responsibilities as soon as I could arrange matters. That I advised him to seek another position to which he might shift as soon as practicable. That in view of his reappointment in Yale for another year he would be entitled to continue his work for such term. (1)

Yerkes noted that Tinklepaugh had twice mentioned the matter to him but "in such a way as to mislead instead of informing."

This was written just three days *after* Yerkes acknowledged receipt of the full information concerning the chimcracker matter. This places in serious doubt Yerkes' contention that the matter was a minor factor in Tinklepaugh's dismissal and any defense based on the early date at which Tinklepaugh was informed that he would be removed.

I found no indication that Tinklepaugh's honesty was ever overtly in question. However, in a June 10 letter to Jacobsen and Nissen, Tinklepaugh clearly felt that it was and wrote "apparently contrary to your opinions, I was entirely honest and frank in my responses to the letters in which you so kindly advised me of the errors that occurred in the article." Tinklepaugh blamed the error on his reliance on data given to him. "The analysis of the chimcracker was made by a graduate student in the Department of Physiological Chemistry, to whom I was referred by Dr. Cowgill" (1932d). In later apologizing for Tinklepaugh, Fulton noted that Tinklepaugh had erred "out of the fact that he trusted implicitly in the analyses which had been done for him by others" (1933a). Tinklepaugh admitted to Angell that he had been careless, but was emphatic that he had not been dishonest (Angell 1933c).

OTHER MATTERS

In addition, Yerkes found fault with Tinklepaugh's handling of the breeding program. Tinklepaugh was convinced that the miscarriages that occurred during these early years were due to excess excitement in the colony. When Yerkes arrived in Orange Park, he became convinced that the problem lay with diet. Yerkes concluded that "whereas the diet which was employed by Dr. Tinklepaugh was adequate for the males of the colony and for immature individuals, it proved inadequate for pregnant and lactating females" (Annual Report 1933, 24). Yerkes added milk, eggs, and cooked beans to the diet and provided continuous access to salt. The result was normal reproduction in the chimpanzees and embarrassment for Tinklepaugh.

There remained some health problems in the colony. Again, Yerkes found Tinklepaugh's procedures wanting. The animals were infested with a variety of parasites and Tinklepaugh had worked to find drugs that would control them. Yerkes (Annual Report 1933, 26) preferred dietary improvement to the use of drugs; the general health and vigor of members of the colony improved. No acute infections appeared and treatment for parasites was unnecessary.

Another apparently minor problem lay in the ordering of chimcrackers. In December 1932, Tinklepaugh (1932b) miscalculated the available supply and the need to reorder. Although there is no paper record of Yerkes' reaction, one can surmise that Yerkes would have been upset with this lack of military efficiency.

Yet another minor source of disagreement between Yerkes and Tinklepaugh concerned the employment of a secretary, Mrs. Elsie Robertson (Tinklepaugh 1932b; Yerkes 1932c). Tinklepaugh was responsible for hiring her, and reached agreements with her, which led her to purchase an automobile to commute from Jacksonville to the station. Yerkes found Mrs. Robertson to be unsatisfactory and she was to be terminated in a manner that Tinklepaugh felt unfair to her. This caused a financial burden on Mrs. Robertson and her newly purchased car. Yerkes stood firm.

Still another incident involved a time when Tinklepaugh wanted to X-ray some chimpanzees in order to study their skeletal growth. Yerkes objected, proposing that the exposure to X-rays might be harmful to the animals and disrupt normal development. Psychologist Heinrich Klüver (Weaver 1937c) believed that this was not the true reason; rather, it was because this was not originally Yerkes' idea. When the work was done and praised at a later time, however, Yerkes was quite willing to take credit for it.

REACTIONS TO THE DISMISSAL

Word of Tinklepaugh's dismissal spread widely. The general impression was that he had been terminated because of the chimcracker episode and that the termination was unfair. Yale University President James Rowland Angell (1933a) wrote to Yerkes that he "was disturbed to learn that your severance of relations with Tinklepaugh has led to rather extreme and unhappy reactions in the Carnegie group in Baltimore." Tinklepaugh's Baltimore advocate seems to have been Carl Hartman, with whom Tinklepaugh coauthored three articles. Angell was concerned that the episode might lead to prejudice against the station. Rightly or wrongly, it was generally perceived that Tinklepaugh had been dropped because of the one incident rather than because of his general unsuitability for the laboratory (Angell 1933b). Yerkes (1933h) even reported experiencing "misunderstandings, slights, and insults that were difficult to bear" from the Jacksonville community. The situation was exacerbated by Bingham's controversial departure just a few years earlier (Angell 1933b, c).

It was the involvement of John F. Fulton, the powerful Chairman of the Department of Physiology, that really upset Yerkes. Fulton mentioned the adverse criticism in Baltimore and added that "at the meeting of the New York Branch of the American Psychological Association last Saturday the comments were outspoken in the extreme and they came to my ears from all sides" (1933a). Fulton warned Yerkes of the widespread perception that a "rank injustice" had been done. Although Fulton agreed that Tinklepaugh may have been careless in making an error of judgment, he noted that Tinklepaugh was a man of integrity and the error was a relatively minor one. Fulton was concerned that the episode might "completely ruin his [Tinklepaugh's] scientific career." Fulton added that it would be difficult for Tinklepaugh to find employment elsewhere because other laboratory directors were concerned that Yerkes, in his powerful position as Chairman of the Committee for Research in Problems of Sex, would retaliate and deny funds to any laboratory employing Tinklepaugh.

Yerkes was outraged, writing to Angell that "it is the most extraordinary letter I have received in my thirty-odd years of administrative and other professional experience" (Yerkes 1933d). Yerkes believed that Fulton's letter "indicates surprisingly little confidence in my integrity, sense of justice, soundness of judgment, or practical experience" (1933h). Yerkes escalated the conflict by proposing that he prepare an article for *Science* magazine explaining his reasons for firing Tinklepaugh. Both Angell and Fulton strongly advised against that approach, as it would hurt Tinklepaugh even more (Angell 1933c, Fulton 1933b).

The situation soon worsened. The dean of the Graduate School and Fulton recommended that Tinklepaugh be awarded a Sterling Fellowship so that he might continue his research with his supervision transferred from Yerkes to Edgar Allen in New Haven (Angell 1933d). The proposal also was made in a letter signed by Raymond Dodge, Fulton, Clark Hull, Walter Miles, and Roswell Angier (Dodge et al. 1933), all well-known and powerful faculty members; clearly Tinklepaugh had strong support among Yerkes' colleagues. Angell favored the appointment but was wisely concerned about Yerkes' reaction. Yerkes indicated that he could not approve the suggestion, as it would "place me in an impossible relationship" (1933f). He felt that he had been "maltreated by his friends" and that "my colleagues in physiology and psychology have clearly enough indicated their lack of confidence in my wisdom, justice, and integrity" (Yerkes 1933g). Yerkes (1933g) continued by resigning his directorship: "Therefore, in the interest of the novel and promising undertaking which we have labored so strenuously to start, and in the service of which during the past three years I have sweated blood, I request that I at once be relieved of administrative responsibility for the Laboratories of Comparative Psychobiology. I should choose to continue for a time in my Yale professorship, unless you deem it necessary or desirable that I resign."

Of course, Yerkes' friend, Angell, would not accept Yerkes' resignation (Angell 1933e). Angell (1933f) wrote a letter designed to mollify Yerkes, pointing out that nobody was questioning his judgment and that these men were his friends. Angell was concerned about the effects of the gossip on both the station and the career of Tinklepaugh. Fulton (1933c) also tried to calm Yerkes.

Angell (1933f) proposed that a fellowship be granted to Tinklepaugh that would enable him to work for another year at a different institution, such as Johns Hopkins. Yerkes (1933b, i) agreed to such an arrangement, provided that the fellowship would not be administered through the Department of Psychology, that it would enable Tinklepaugh to complete writing up data already collected, and that it would be at another institution.

TINKLEPAUGH'S LATER CAREER

It appears as though the proposed fellowship for Tinklepaugh was never activated. He left Yale, and the academic life, in 1933. The rest of his life would be spent in the business world.

PERSPECTIVE

How are we to evaluate these events? Was Yerkes justified in firing Tinklepaugh or was the dismissal just one more instance of Yerkes' alleged tendency to purge independent thinkers? Surely, Yerkes was closer to the situation than were his critics. We do not know all of the details that were involved. Perhaps Yerkes' critics did not know them all either. Yerkes may have been correct in questioning Tinklepaugh's overall suitability as a long-term staff member.

On the other hand, Tinklepaugh gained the support of a considerable number of important scientists both at Yale (e.g., Angier, Angell, Dodge, Fulton, Hull, and Miles) and elsewhere (e.g., Corner, Hartman, and Tolman). Even as he was questioning Yerkes' scientific ability during the 1937 crisis at the Rockefeller Foundation, and several years after Tinklepaugh left animal research, George Corner (1937) wrote to the Rockefeller officials that he thought well of Tinklepaugh. Max Mason (1928b) thought that Tinklepaugh seemed able. Was this all due to personal charm or was he respected as a scientist? If he was not creative, at least he was more productive than most of the other members of the scientific staff. In contrast to Tinklepaugh's nineteen articles, Yerkes would have trouble getting most of his staff to complete and write up research.

At the least, Tinklepaugh appears to have been an active and competent scientist. Yerkes comes off, by contrast, as one who was easily offended when his authority was questioned, even in the interest of a rather moderate compromise. He seems to have deliberately falsified the reason for Tinklepaugh's dismissal. The timing of the dismissal and Yerkes' letter to Nissen and Jacobsen strongly suggest that the chimcracker caper was the primary reason for it. These events forced Tinklepaugh out of the academic world and contributed to Yerkes' reputation and later problems as well. Thus, both were hurt by the incident.

Other Departures

Others who left the YLPB had better experiences with Yerkes. Haslerud (1977) recalled that when he impulsively accepted a job at the University of Tennessee, Yerkes "angry and disappointed . . . denounced my irresponsible decision but then ended up with good wishes for my success in the academic position" (26). Similarly, Meredith Crawford (1977) recalled a pleasant departure from Orange Park. Surely the psychologist working at Orange Park in the 1930s who was most independent of Yerkes was Kenneth W. Spence. However, there is no indication of strife between them.

It is likely that some departures were justified when staff members failed to live up to expectations, although it is difficult in retrospect to separate these from the difficulties discussed above. The failures of Joseph Yoshioka to write up his research and of Michael Tomilin to complete his doctoral studies appear to have justified their dismissal.

Near the end of Yerkes' administrative tenure, the quiet Henry Nissen expressed a balanced view of Yerkes' relationships with his staff to Rockefeller officer Frank B. Hanson. According to Hanson,

> Y[erkes]' treatment of former staff members at Orange Park leaves much room for criticism and complaint. However, this being true, other things for which Y[erkes] was not responsible are debited against him. . . . He [Nissen] believes that some of the men Y[erkes] brought here were clear misfits, and that a more discriminating director would never have appointed them. One of Y[erkes]' weaknesses as an administrator was that he allowed a bad situation to develop to the explosion point before taking action, at which time, however, he took too drastic action. The latter seems to be the case with both Tinklepaugh and Bingham. (Hanson 1939)

Life in Orange Park

Although Orange Park today is a bustling, suburban community, in the 1930s it was a small, out-of-the-way locale, especially in the eyes of some of the northerners from Yale. With housing in short supply, some investigators had to live, with some inconvenience, in Jacksonville; there was no adequate bus or train service. Many of the amenities taken for granted in New Haven were missing in Orange Park.

A house on the riverfront was leased for use of the director when he was in residence and his deputy, often Elder, when he was not there (A. W. Yerkes 1939). It had an old swimming pool that provided some recreation for the staff. Other staff members rented various homes around Orange Park when they could.

In general, one gets the impression that the laboratory group was close and that they got along well with the local residents. As former secretary Marcia Hewey (Anderson) confirmed, the laboratory employees were especially welcome because they contributed to the economy of the small town by renting homes, patronizing local merchants, sending children to local schools, being active in local affairs, and employing some residents (Anderson 1993).

Perceptions from the Community

Yerkes believed that the animals needed quiet. Therefore, while he was open to visiting scientists and loved to show off his facility, he kept nonprofessionals away from it. This led to many misperceptions of the activities of the station. One view was that Yerkes was trying to influence the local elections by bringing in a group of Yankees whose votes he could control (Coolidge 1977). Another popular view was that the Yale doctors were trying to crossbreed the chimpanzee with humans, or "crossbreeding with the 'colored' " (Nowlis 1994a, 3) behind those barbed-wire fences.

Reflections

Many staff members and visitors reflected on various aspects of the YLPB in interviews and other venues.

Animal Anecdotes

Asking veterans of the Orange Park days for reminiscences usually included two kinds of stories: those of occasional escapes by the chimpanzees and those of chimpanzees either spitting water or flinging feces at passersby. Nowlis (1994a) recalled a case when a friend of theirs, well groomed and stylishly dressed, was able to visit the laboratory: "Escorted by Austin [Riesen] and me Mary advanced to a spot about 15' in front of a line of four outdoor cages occupied by a total of perhaps five females and one male, all adult. The effect on the females was quick and dramatic. Their two old, dependable friends had appeared with a highly decorated and adorned female with dominating posture and stare! Mary became the hated target of shouts, screams, rattling of the walls and of hurled feces, vomitus and other material. Our departure was silent and prompt" (37–38).

The chimpanzees could carefully remove a thorn from a human's finger (Haslerud 1977). Many researchers recalled how each animal had a different personality. Some chimpanzees were extremely cooperative; Alpha, by contrast, would bite Crawford when in an experiment, so he had to carry a small rubber hose, with which he could hit her just hard enough to end the encounter. Nowlis (1994a) recalled that some chimpanzees seemed to enjoy the summer thunderstorms. A young male, Frank, responded as if to an unseen rival, strutting, stomping, and beating his chest.

New Haven

Several former staff members, including Jackson (1993) and Spragg (1993), commented on the intellectual excitement of Yale psychology in the 1930s. Haslerud recalled that "the very air was electric with ideas and intellectual stimulation" (1977, 21). Both Haslerud and Meredith Crawford (1977) recalled inexpensive and stimulating lunches in the "blue room" at Yale. There were also the evening seminars of Hull, Yerkes, and others.

Orange Park

Orange Park in those days provided quite a contrast. Haslerud (1977) described the culture shock he experienced upon arriving in Orange Park in his secondhand $100 Model A Ford. He recalled that the drinking water was heavily impregnated with hydrogen sulfide. The insects soon gave him problems with skin infections. There were scrawny cattle and razorback hogs roaming about under Florida's open-range law. The two-mile road to the laboratory was unimproved. The boom calls of alligator bulls in the evening provided reminders of the environment in which he lived. Nowlis (1994a) recalled the incredible heat of the Florida summer in the days before air conditioning and the huge populations of cockroaches in most homes and especially at the Laboratories. Crawford (1993) remembered that there was one store, a grocery store, that had meat that was generally quite tender—because it often was covered with a green mold.

As Ada Yerkes was writing *Yale in Florida,* she was distressed by several signs of "progress" in Orange Park. For fifty years prior to 1939, a triple row of old oak trees occupied the center line of Kingsley Avenue, the central street in Orange Park. Botanist Ada Yerkes (1939) lamented the 1939 decision to fell those trees in the name of progress. She even lamented the coming of dial telephones.

The Staff and Its Pastimes

Most agreed that the staff was tight-knit and often socialized among themselves with little interaction outside of laboratory personnel (Spragg 1993). An exception was Helen Riesen, who recalled greater interactions with her neighbors (Riesen and Riesen 1994).

Haslerud (1977) recalled evenings in the homes of the various researchers; there was little other entertainment available. Evenings in the Yerkes home often would be associated with the parlor games beloved by the family. There would be no alcohol at the Yerkes'. Mrs. Yerkes ran a sewing circle which all of the laboratory wives were expected to attend (Crawford 1993).

There was much outdoor life in the warm Florida climate. Both the

Finches and the Crawfords obtained sailboats, and boating became a major source of recreation. Nowlis (1994a) remembered that fishing was popular. Graham Townsend, who would be the second superintendent, was an excellent fisherman, having fished the St. Johns River all of his life. Of course, swimming in the ocean and sunning, shelling, walking, and bird-watching on the shore were popular pastimes. The hard-packed beaches of the area were much more sparsely inhabited in those days. Elder and Crawford would saw and chop firewood on the laboratories' property. Spence and Crawford would play first-rate tennis (Crawford 1993).

Crawford (1993) recalled trips to Jacksonville for shopping and movies. Another diversion was provided by trips to the county seat, Green Cove Springs, for mineral-salt massages in the pool of sulfurous water (Nowlis 1994a). Nowlis recalled the classical music concerts in Jacksonville.

The flow of distinguished visitors to the laboratories provided some scientific excitement (Haslerud 1977). The visits of the Scientific Advisory Board were viewed as opportunities for helpful advice.

Robert Yerkes

Of course, there were many comments on Robert Yerkes himself. The descriptions were generally consistent with those of others. Helen Riesen recalled Robert Yerkes as being very charming, but quite formal (Riesen and Riesen 1994). Spragg (1993) recalled Yerkes as "a rather shy, remote individual" and "quite formal in personal relationships." Nowlis noted that Yerkes was a man of precise habits; "you could almost set one's watch by his leaving his office with Helen Morford and walking down the hall and going to lunch, driving his Buick, etc. and then coming back" (1994b).

While disembarking from his first sailing trip with the Crawfords, Yerkes managed to fall overboard, providing the basis for many stories in the lore of Orange Park (A. W. Yerkes 1939). There seem to be several versions to this legend. As told by Crawford (1993), as the boat approached the dock, Yerkes tried to jump to it. As each action has a reaction, the boat moved back and Yerkes wound up in water over his head. Yerkes went home, while the Crawfords continued sailing. When they later checked in on him, they found that Yerkes had told his wife that Crawford had capsized the boat. About three days later, the Yerkes came for a crab dinner at the Crawfords. All got sick from the crab. By the next APA meeting, the joke had gotten around that the Crawfords had tried twice to kill Yerkes—first by trying to drown him and later to poison him.

Then there was the Fernandina Oyster Bar, where one could get all you can eat for one dollar: "Yerkes, always restrained and abstemious, is reported to have once eaten 12 dozen of the relatively small oysters, and sailed all the way back to Orange Park only to end a perfect day by having to be carried off the sailboat" (Nowlis 1994a, 24).

David Yerkes (1993) saw Robert Yerkes in the more casual, family setting. He recalled that his father was effective at raising funds for his various projects, but not because he was a dramatic talker. Rather, he was firmly convinced of his ideas and that message came across. David described his father as having a sense of humor, but more responsive than creative and never the "life of the party." He enjoyed friends, games, and the humor of others. He remembered his father as rather pedantic and ponderous in his writing style, but more easy and informal in conversation.

James Harlan Elder

The most puzzling recollections are those of James H. Elder. In his Annual Report of 1936, Yerkes praised Elder, noting, for example, "the intelligence and efficiency with which Doctors Nissen and Elder have performed their administrative duties" (20). This situation then changed so much that by 1939 Yerkes commented that "it is not deemed desirable that Doctor Elder's services in his present capacity be continued longer than necessary. He will be advised to seek appointment elsewhere" (Minutes of Meeting of Administrative Board, January, 1939, 6). Something had changed. In August 1939, Elder accepted an appointment at the University of Wisconsin.

Elder (1977), in a later talk, was generally positive concerning his experiences at the YLPB. By contrast, in a remarkable, unpublished 1939 manuscript, he presented a very different view of Yerkes' administrative abilities and style. Elder began with a statement of his goal: "This heretical venture will attempt to prove involuntary mismanagement and misdirection of the Yale Laboratories of Primate Biology during the past ten years. It will attempt to show that crucial decisions were made, not on the basis of objective data, but in reference to an individual's whims. Consequently, many decisions were openly contradictory to the facts. The most serious charge which will be made is that research is determined and almost always dominated by a unique personality, frequently at the expense of science" (Elder 1939a, 1–2).

Elder contended that, contrary to Yerkes' initial proposals for the station, nonpsychologists had been unwelcome. He attributed this to Yerkes' alleged "lack of ability to supervise work of any other kind" (2). Elder cited Yerkes' "bungling of the drug addiction research—progressing nicely in the hands of Doctor Spragg—and his failure to comprehend certain fundamental principles of reproductive physiology as conspicuous examples" (2).

Elder (1937b) had written to Yerkes, pointing out that the files now contained such rich longitudinal material and that an expert be brought in to ensure that the tedious measurements being conducted were worthwhile. Yerkes (1937f) brushed off both that suggestion and one by Nissen (1938) that a social psychologist and a young physiologist, anatomist, or zoologist be added to the staff.

Elder (1939a) believed that the reasons for the narrowness of approach lay in Yerkes' personality. Yerkes resisted research where there would be "the difficulty of actively 'directing' and dominating these men and their work" (4). What were critical, he contended, were "a minimum of independence and a full measure of humility" (6) on the part of the staff.

Elder described a pattern in which Yerkes intimidated his staff as they prepared written reports. If an investigator prepared the report, he "submits it to scathing criticism of his English" (9). This may have been one reason for the staff's reluctance to complete such reports. "If, however, he accepts a minor responsibility in publication he may find himself in the embarrassing position of being forced to lend his name to statements which he knows are in violation of scientific truth" (9). "The impertinence of original criticism does not lead at once to dismissal, but the offender is tagged as one who is incapable of grasping the spirit of cooperation or as one who is unable to think clearly and do good research" (9). Thus, "men rise and fall in the Director's estimation not on the criteria of research ability but because of a change in the Director's relationship to that research" (12). In support of the charge that one sometimes had one's name placed on conclusions one could not support, Elder cited his own experience with work on dominance and the menstrual cycle. He believed that Yerkes' then recent paper (Yerkes 1939b) was a rehash of earlier data and that "in both [papers] he presents conclusions which are not supported by the data" (18). In one case, Elder accused Yerkes of having "to change the record several months later, in order that it might conform to the facts!" (19) (see chapter 5).

Elder wrote of Yerkes' "emotional orgy of constructing new buildings" (12). He noted that Yerkes complained about having to take time from research to worry about building buildings, when, in fact, it was buildings that were his passion (15).

Elder contended that there was much information in a recent *Scientific American* article (Newman 1939), published with Yerkes' approval, that was inaccurate and misleading. (My examination of that article reveals several misleading statements but suggests that Elder may have overstated his criticisms.)

He got personal. "Many people are inclined to regard the Director as a modest, unobtrusive, little gentleman who has unselfishly dedicated his life to Science" (14). "Those who work closely with him soon discover that almost every action reflects the extreme vanity which completely dominates his life" (14).

Does this manuscript simply reflect the sour grapes of a disgruntled former employee or is it an accurate portrayal of life at the YLPB? It matches material from some sources but is inconsistent with the views of some colleagues (e.g., Meredith Crawford and Vincent Nowlis). Elder admitted some pique as motivation for writing the piece and defended his work at some length. As I can determine, Elder made some important and valid points, as confirmed, for example, in the RF reports of 1937, but these are mixed with some hyperbole as well.

7
Retrenching

The 1935 Attempts at Restructuring

By 1935, the basic feasibility of breeding, maintaining, and studying captive chimpanzees had been established. A crossroad had been reached and it was time to step back and reevaluate the project that had been brought to fruition. In one way or another, that reevaluation would continue for the rest of Robert Yerkes' tenure with the YLPB.

Incorporation

The process began with the meeting of the Board of Scientific Advisors in Orange Park in March 1935. It had become apparent that the Anthropoid Experiment Station had been operating illegally in the State of Florida. Advice from the Florida Attorney General Landis and Jacksonville attorney Perse L. Gaskins confirmed that the station was liable to taxation in the State of Florida, in spite of the research and educational missions, because it was owned and operated by a "foreign" corporation, Yale University. Indeed, such foreign corporations were prohibited from owning property or transacting business in the State. Several possible solutions were considered. The most obvious solution was for Yale to establish a nonprofit corporation within the State of Florida. Yerkes proposed this to the Prudential Committee of Yale University, which acted favorably.

Thus, as the result of action of June 11, 1935 in the court of Judge DeWitt T. Gray, the property was transferred to a new corporation, the "Yale Laboratories of Primate Biology, Incorporated." The new corporation was composed of the members of the Prudential Committee of Yale University and several fiscally responsible officers of the New Haven institution (Annual Report 1935; Charter 1935).

The new name, the "Yale Laboratories of Primate Biology" replacing "the Laboratories of Comparative Psychobiology, Yale University" was significant, in that Yerkes wanted to leave open the possibility of studying a range of primates species, including humans and, more importantly, he wanted to pave the way for an expanded mandate that would take the fields of investigation beyond the limits of "psychobiology" (Conklin, Lillie, and Yerkes 1935).

An Attempt at Restructuring

The 1935 board meeting also addressed the issue of the status and future plans for the station (Minutes of the Scientific Advisory Board March 4 and 5, 1935). There was uncertainty about how far and in which directions research at the facility could move within the limitations of finances and staff that existed at that time. At one extreme, the possibility of liquidation was considered. That rejected, various forms of reorganization were considered. As anyone experienced in committee work might predict, the outcome was to approve a motion that a subcommittee would be appointed to study the matter further.

The subcommittee (Conklin, Yerkes, and Lillie) produced a thirty-eight-page report, authored primarily by Yerkes (Conklin et al. 1935). The group proposed a new, and rather grandiose, ten-year plan with an increased budget, a concentration on a few major research areas, and major appointments of scientists to three-four research professorships in addition to a bevy of eleven-fourteen Assistant Professors, Associate Professors, Instructors, and Assistants. The report was considered in New Haven and at the RF but it was never acted upon.

The Crisis of 1937

Events at the YLPB cannot be understood without a thorough examination of the interactions between Yerkes and the Rockefeller Foundation (RF) officers in 1937. These events in some sense were the defining events of Yerkes' tenure because they shaped the future of the facility that Yerkes had worked so long and hard to develop. The initial RF appropriation was due to expire June 30, 1939. Yerkes had been pushing for the implementation of his new ten-year plan that would entail expansion and increased funding. A decision had to be made regarding continuation or expansion of RF funding. The negotiations would not go well, with the ultimate effect being that the only way to save the facility was for Yerkes himself to step aside.

Intimations of Difficulties

Even as the RF was providing funds to found the station, the comments of some of its officers show that Yerkes was regarded as a somewhat marginal character. One gets the impression that some officers, especially Warren Weaver, regarded him as a bit of a nuisance. Later, the officers had become skeptical of his requests for special funding when they felt that he had not yet adequately demonstrated his ability to use the funds already allocated to maximal benefit.

Relationships with the RF officers soured further when Weaver and Lawrence K. Frank visited Orange Park early in 1935. In 1929, Frank, who was then the Executive Secretary at the Macy Foundation, had represented the Memorial in negotiating arrangements for the founding of the Station; however, his first opportunity to visit the facility came that February. In his diary entry, Frank (1935) raised what would become one of the central issues—the need for broad utilization of the Orange Park facility. Yerkes was committed to the notion that the facility could best be utilized with a full-time, paid research staff. Frank saw this as more a product of Yerkes' rigid, personal beliefs than as based on any actual difficulties that Yerkes had experienced. One might note accusations of Yerkes' desire to control his staff tightly. Frank thought that the station should receive the broadest possible use, as suggested in various memoranda from Yerkes when the station had been proposed. His models were the Marine Biological Laboratory at Woods Hole and the Naples Zoological Station. Frank granted Yerkes' accomplishment in demonstrating the feasibility of breeding and maintaining a chimpanzee colony, but envisaged leading biomedical scientists and experts in child development traveling to Orange Park to make maximum use of the facility in a wide range of fields and to bring expertise to the station that would improve the quality of the work there.

Weaver, who had become the Director of the Division of the Natural Sciences at the RF, the unit then responsible for funding the YLPB, visited in March of the same year. Surely, Yerkes saw this as his chance to really impress the key man at the RF and ensure the increased funding he sought for the reorganization of the laboratories. The tone of his diary entry (Weaver 1935) suggests that he regarded Yerkes as a rather tedious host, who could stretch a short point into an afternoon of discourse. The two men never really got along. While Yerkes presented his plan for reorganization and increased funding, Weaver stressed the need for economy at the RF during difficult times when cutbacks, not increases, were the order of the day. Two years later (Weaver 1937b), Yerkes told Weaver that he believed that Weaver's personal antagonism toward him was interfering with the progress of the station and suggested that it might be better for the station if he (Yerkes) stepped aside as director. Weaver denied that there was any negative personal relationship between them. He explained that Yerkes' personal characteristics were relevant to the future of the YLPB only to the extent that they affected its activities but not as the result of any personal antagonism between them.

Weaver (1937a) also introduced another issue that tended to incite Yerkes' ire. The RF provided primary support for the Institute of Human Relations at Yale. Weaver regarded the YLPB as part of that operation and thus felt that the Foundation was being milked twice over as both sought funding. Some of the YLPB monies came from the IHR and thus the total amont of RF support, both direct and indirect, was more than it appeared on paper. For his part, Yerkes contended that the IHR contribution was merely the manner in which Yale was best able to fulfill the commitments it had made to him in the days of the now-abandoned Institute of Psychology and that the station was now virtually completely independent of the IHR.

Getting Expert Opinion

During late 1936 and early 1937, the RF officials sought the opinions of some experts in the field. The results were devastating for the future of the YLPB as Yerkes envisaged it. The officials selected an interesting set of evaluators. Edward C. Tolman, of the University of California, Berkeley, was among the leading animal psychologists of the day. He had sent Tinklepaugh and Yoshioka, both later dismissed by Yerkes, to the YLPB. Solly Zuckerman, a distinguished British scientist, had considered an appointment at the facility, but decided it would be unworkable when he visited Yerkes and found their personalities incompatible. Karl Lashley was a leading physiological psychologist who would succeed Yerkes as director in 1942; Jacobsen and Yoshioka had been his students. William H. Taliaferro, a zoologist and dean at the University of Chicago was a close friend of Lashley. Embryologist/physiologist Carl G. Hartman, of Johns Hopkins, had been critical of Yerkes' handling of the Tinklepaugh situation. Heinrich Klüver was a leading physiological psychologist at the University of Chicago. George W. Corner was an anatomist at the University of Rochester. All except Tolman were committed to more anatomical and physiological methods than to the integrative psychobiology of Robert Yerkes. The cards appeared stacked against the director.

Many of the impressions presented by the reviewers are summarized in the diary entries of Warren Weaver and other officers. The group was generally critical of key aspects of Yerkes and his operation. It is remarkable how similar were the opinions expressed by different scientists around the country. This may have been, at least in part, because of communication among them. Corner for example, noted "my own contacts with him have not brought out any lack of tact or fairness, but I hear rumors to the contrary" (1937). It may also be due, in part, to the RF officers leading their evaluators by presenting them with a set of specific queries.

CHIMPANZEES AS LABORATORY ANIMALS

One concern was that chimpanzees are not ideal animals for many research purposes. They are expensive and difficult to keep. Their reproductive rates tend to be low. Invasive procedures involving surgery and other manipulations can more effectively be carried out on monkeys, which are generally regarded as less valuable. Issues of the ethics of research on chimpanzees were rarely mentioned.

Lashley (1937) provided a detailed analysis of the costs and benefits of research with chimpanzees. He concluded that "from the standpoint of the comparative neurologist and psychologist, the further development of the chimpanzee as a laboratory animal is not of sufficient importance to justify the cost." Hartman, Klüver and Zuckerman agreed (Klüver 1937; Miller 1936; Weaver, 1937a, 1937c, 1937e)

PRODUCTIVITY OF THE LABORATORY

Tolman indicated that Yerkes' output was "far below his opportunities and financial support" (Hanson 1936). Lashley (1937) concluded that "the research of the anthropoid station has been disappointing." Corner, however, was more favorably impressed with the progress at the laboratories than were some of the others.

THE QUALITY OF THE YLPB STAFF

As had Weaver earlier, Kluver (Weaver 1937c) expressed a low opinion of Yerkes' staff. Lashley (1937) added "I think it unlikely that the present staff will contribute anything of fundamental importance here." He evaluated staff members individually, giving generally favorable opinions of Jacobsen, Elder, Spence, Spragg, Tinklepaugh, and Carpenter. He was critical of Nissen as being relatively unproductive and unoriginal, Jacobsen as being specialized in his interests, Spence as having minimal background for work on problems of complex behavior, and Bingham as second-rate. Corner (1937) thought highly of Tinklepaugh.

Hartman thought Nissen better than Yerkes and that he ought to be appointed as the Director (Weaver 1937e). Hartman rated Elder as "mediocre"; Jacobsen "a little bit too enthusiastic about Jacobsen"; and Tinklepaugh as "nearly first-rate." Hartman added that after Bingham had done a good job early, he "went completely to pieces, and H[artman] questions whether this was Y[erkes]' fault." These evaluations are especially interesting in light of Yerkes' own evaluations of his staff discussed earlier.

THE SCIENTIFIC ADVISORY BOARD

Taliaferro (Weaver 1936) was critical of the men who served on the Board: in his view Merriam was not particularly competent, Wells had aged to the point that his judgment was questionable, Conklin was not the man to ask and Lillie was influenced by the early judgments of Yerkes' promise.

YERKES' EXCLUSIONARY ATTITUDE TOWARD VISITING SCIENTISTS

Some of the evaluators agreed with the RF that the YLPB could be used widely by a broad range of scientists. Hartman thought that Yerkes "and the rest of the Station get the jitters whenever visitors arrive" (Weaver 1937e). Frank, Klüver, and Taliaferro agreed (Weaver 1936, 1937c, 1937e).

In a rare exception to the general unanimity of opinion, Lashley and Corner agreed with Yerkes' exclusionary posture. Lashley pointed out that with the present colony size and staff organization, "it is clearly impossible to receive transient investigators, who have little experience in handling the animals, without seriously disturbing the work of the staff" (Lashley 1937).

OVEREMPHASIS ON PSYCHOLOGY

Both Hartman and Frank thought that the program at the YLPB placed too much emphasis on problems of psychology whereas much could be accomplished with more attention to work in other disciplines (Weaver 1937e).

YERKES' TREATMENT OF HIS JUNIOR STAFF

The Bingham and Tinklepaugh cases continued to haunt Yerkes. At the same time that they criticized the quality of the YLPB staff, the evaluators criticized Yerkes for dismissing those staff members whom he had released.

There was a general perception, here as elsewhere, that Yerkes tended to dominate his staff, demand their lockstep adherence to his program, and dismiss those who would not fall into step. Tolman summarized the majority view that had spread around the country that Yerkes was "adept in getting rid of his best young men, magnifying their minor faults and overlooking their real values" (Hanson 1936). Klüver pointed to the tragedies of Bingham, Tinklepaugh, and Yoshioka, all of whom were struggling in one way or another at the time (Weaver 1937c). Lashley (1937) echoed the general theme: "in selecting and handling men he [Yerkes] seems . . . [to prefer] men who fit the routine of the general project and are willing to follow his lead in research. The more original and independent men tend to break away."

YERKES AS AN OLD-FASHIONED NATURALIST

There was a feeling that Yerkes was too much of an old-fashioned naturalist in an era when the modern approach was becoming more physiological and reductionistic. Klüver argued that Yerkes was primarily a naturalist who "works on chimpanzees, not on problems" (Weaver 1937c). Lashley (1937) wrote that Yerkes "has tended to descriptive and naturalistic studies rather than to the experimental analysis of complex functions." Even Tolman agreed: "Y[erkes] is fundamentally a naturalist . . . [who] thinks he is being a scientist when he records that a certain experiment was performed on [*sic*] 10:15 A.M., Tuesday, January 13" (Weaver 1937d).

YERKES' AGE

The question of Yerkes' age, 60, was raised by Taliaferro (Weaver 1936). Tolman was "emphatic that 'old men' should not schedule the development nor dictate the problems of such a project, but should welcome the vision and imagination of young men" (Weaver 1937d).

ROBERT YERKES AS A MAN AND SCIENTIST

Yerkes lacked the respect of his evaluators. Zuckerman made "some rather devastating criticisms of Y[erkes] as a person and as a scientist" (Miller 1936). Taliaferro expressed "some reservations concerning Yerkes' capacities for effective leadership and development of such a project, and perhaps even of his own ability" (Weaver 1936). Taliaferro

(Weaver 1936) thought Yerkes "was a man of originally great promise" but that he had little to show for his support. Klüver thought that there were important problems to be studied with chimpanzees, but that Yerkes "has neither the training, ability, nor desire to work on them" (Weaver 1937c). Klüver suggested that Calvin Stone could have done a much better job with the facilities than could Yerkes. Corner (1937) concluded that "I do not think Dr. Yerkes is to be regarded as a scientific investigator of the first rank" and rated his common sense as average.

Hartman revealed that Yerkes had once asked him what other people thought of him. Hartman replied to Yerkes frankly that: "a good many of his scientific colleagues had a very low impression of Y[erkes] as an individual . . . he did not know a single person who had ever worked with him or for him, who liked him . . . he ought to be a good enough psychologist to know that persons cannot be treated as if they were machines" (Weaver 1937a). Hartman was impressed with the "sportsmanlike and honest way" in which Yerkes reacted to this harsh assessment. Hartman later praised Yerkes' efforts in establishing the facility, noting that "his single-handed efforts in bringing the anthropoids into their own constitutes a milestone in biological history" (Weaver 1937e).

The harshest judgment came from Lawrence K. Frank, whom Weaver (1937f) quoted as thinking that "any discussion of this project must begin with the recognition of the fact that Yerkes is mentally ill." Frank felt the laboratories worthy of continued support, but only if Yerkes was replaced.

OVERALL OPINIONS

The consensus was that the station should be continued, but without increased support and without Yerkes. Tolman (Weaver 1937d) believed the project "so important that it would be terrible if it were dropped." Klüver also thought the project of sufficient importance to warrant its continuation. Corner thought the project important, though other programs might outrank it. Hartman argued for the continuation, but not expansion, of the project. Lashley expressed an unfavorable opinion of the work at the station. Overall, there was little enthusiasm for expanding the project beyond its then-existent scope as Yerkes had proposed.

EVALUATING THE EVALUATIONS

Although some of this advice seems, with the benefit of time, to have been reasonable, other suggestions were less so. Klüver's and Lashley's belief that problems of learning and performance in complex tasks could be conducted as effectively with monkeys as with chimpanzees has been proven simply incorrect with recent work such as that of workers such as Rumbaugh, Savage-Rumbaugh, Premack, Gallup, Povinelli, Gardner and Gardner, Tomasello, and Boysen.

Despite his human foibles, I find little evidence to support Frank's view that Yerkes was mentally ill. He was not at home in social situations, and this may account for his failure to understand the changes occurring around him that would eventually lead to his departure.

Yerkes did have some problems dealing with his staff. The view that Yerkes needed to dominate and control their work seems real. He was caught in a bind, being criticized for both releasing staff with whom he had problems and retaining staff whom the evaluators regarded as of marginal competence. The Tinklepaugh and Bingham cases were more complex than was realized by the evaluators. Perhaps the truth lies in between the extremes in that Yerkes had serious staff problems but his handling of them was not quite as bad as was perceived at the time.

It would seem as though Yerkes could have opened the station to more investigators, but the solid database of detailed information on the growth and reproduction of these animals was important and worthwhile. With more funding, it might have been possible to add chimpanzees from other suppliers for the use of visiting investigators, as suggested by Hartman, without disrupting the work on the core colony. One must have some sympathy with Yerkes' desire to generate solid, normative data in a stable environment.

The field was moving in a more reductionistic direction at the time. It would swing back after World War II, when the full impact of the work of European ethologists helped to enhance new perspectives in animal behavior. Today, many comparative psychologists would regard Yerkes' naturalistic approach as a strength rather than as a weakness.

Toward Resolution: Yerkes Considers Leaving

Repeatedly, though somewhat inconsistently, through this process, Yerkes considered the possibility that the station might do better if he left as its director. Eventually, it became fully apparent that the only way in which he could save the station was to withdraw. In early 1937, he suggested to Weaver the possibility that he might step aside. In May, Yerkes (1937c) wrote to Angell that "my only way to assist further, and perhaps even to save the Laboratories from wreckage, is to withdraw from the leadership." He added that he had discussed the possibility with Lillie, who thought it a good idea. He wrote to Alan Gregg, announcing his decision to withdraw, blaming "Dr. Weaver's attitude and actions" (Yerkes 1937d). Yerkes, as might be expected, sounded depressed: "They have downed me at last, thanks chiefly to destructive criticisms."

Weaver (1937b) was concerned that Yerkes apparently had in mind only a partial withdrawal that would make it difficult for his successor to effect any really major changes. In fact, Yerkes later vacillated and became rather bitter about his treatment. In the end, he made a phased withdrawal.

The times were changing and Yerkes was caught up in the middle of the changes. Regarding the attitude in New York, Yerkes (1937c) wrote to Angell that "within a few months, it has turned from white to black." He was puzzled over the dramatic change that occurred, when two years ago favorable attitudes seemed universal. Yerkes had been so much in touch and in harmony with the developments around him during the 1920s that he was able to secure support for the station. Now, just a few years later, he seemed out of touch and without support. In the midst of other changes, his longtime friend and supporter, President Angell, retired in 1937, to be succeeded by Charles Seymour, who would be supportive of the station but less so of Robert Yerkes.

Resolution

Onto the scene rode the knight in shining armor who would save the laboratory, but only through the sacrifice of its founder. Stanhope Bayne-Jones was a bacteriologist and a Yale alumnus with a Johns Hopkins MD degree. As the dean of the medical school at Yale, he understood the situation and took effective action to, in effect, save the YLPB. The process began with a visit to Weaver at the RF (Weaver 1937g). Bayne-Jones not only understood most of the criticisms of Yerkes and the YLPB, but he agreed with them. Perhaps it is not surprising that a bacteriologist/physician would adopt an approach sympathetic to the reductionistic emphases emerging at the RF and elsewhere in opposition to the more integrative approach represented by Yerkes. At any rate, Bayne-Jones agreed that Yerkes' plan was too pretentious, the work was too naturalistic, and that the emphasis was on apes rather than problems. Chimpanzees were not being used for purposes that were most appropriate, material should be used more widely, resignations had decimated the staff, and "that some of these faults cannot be expected to be remedied under Y[erkes]' leadership." However, he stressed that Yale remained vitally interested in and committed to the project. Bayne-Jones's proposal to Weaver was that President Seymour appoint a committee of Yale faculty to consider the situation and make recommendations.

THE COMMITTEE ON THE YALE LABORATORIES OF PRIMATE BIOLOGY

A committee of Yale faculty members was duly established and included Bayne-Jones as chair, with Provost Edgar S. Furniss, physiologist John F. Fulton, anatomist Edgar Allen, and psychologist Walter R. Miles, who served as secretary. As the committee held its first three meetings it also consulted with a variety of outside evaluators, including some who had advised the RF. The conference affirmed the many contributions of Robert Yerkes and the YLPB, but repeated many of the criticisms as well. The overall tone of the meeting, on the turf of Yale, to the extent that one can judge from its minutes, was more restrained than were the earlier opinions given to the RF officials (Bayne-Jones 1937). The major new criticism that was recorded was that the physical development of the chimpanzees may have been retarded somewhat because of the failure to provide larger cages and adequate opportunity for exercise.

The group recommended the continuation of the YLPB, with an increase in the number of species kept. In order to be closer to Yale and its facilities, the committee thought that the facility might best be moved to New Haven, however. It was generally accepted that Yerkes had agreed to step down as director; Carlyle Jacobsen was the group's choice to succeed Yerkes. At the same time, the group members wished to ensure a workable situation in New Haven for Yerkes. Knowing Yerkes' sensitivity in such matters, they drafted and passed a resolution affirming the importance of Yerkes' accomplishments with the station.

Yerkes (1937b), meanwhile, had written to President Seymour with his views on the situation. He had some suggestions regarding the future of the station and wanted himself and his New Haven operation integrated within the Department of Psychology. At the committee's fourth meeting, however, it was apparent that while the department would welcome interactions with Yerkes, Chairman Roswell P. Angier did not want him as a member. Yerkes at this point was at risk of becoming a man without a country, as he would withdraw from Orange Park but was unwelcome in the Department of Psychology. The problem with his appointment and salary was that Yerkes had originally been appointed in the Institute of Psychology and later moved to the station budget, funded through the RF. He seemed unwelcome in any of the existing administrative units. Although the Yale administrators felt an obligation to continue his appointment, they did not wish to allocate additional funds to pay for his salary and could see no appropriate unit within which to do so even if could.

The committee met with Yerkes on January 27, 1938. Bayne-Jones opened the meeting with a brief summary of recent events, including the fact that Yerkes did not care to continue as director. Yerkes emphatically corrected that view: "I would put it that because of developments I felt I was forced to withdraw in the interests of the project. This seemed necessary to relieve the University and the Rockefeller Foundation of embarrassment. Willingness to sacrifice myself for the benefit of the project is quite different from any desire on my part to withdraw from the directorship" (Miles 1938a, 1). Yerkes' hurt, expressed in several places, was clear. For example, in 1938 he would reflect on "the deluge of destructive criticism and unpleasant looks which have come to me, even from some whom I thought my friends" (Yerkes 1938).

Yerkes was not done, however; he came back fighting. He had some strong views on a future director. He felt that he could not recommend Nissen. He was even more emphatic about Jacobsen, whom he had earlier thought of as his possible, eventual replacement. Yerkes thought Jacobsen's line of research, on the neural bases of primate behavior, was too specialized and noted that he and Jacobsen had fundamental, mutual differences in point of view. Further, Yerkes asserted that "the appointment of Jacobsen would be viewed by the scientific world as an expression of disapproval of the past and present work of these Laboratories and that this action would reflect unfavorably on the University" (1). One can recognize the same rhetorical strategy that Yerkes had used in the matter of Tinklepaugh. He went further, "If the committee feels that my services should be disapproved publicly, then I shall stand aside in conformity with my statement of willingness to withdraw. However, I would now be willing to recall my proposal that I be relieved of the directorship, if this would solve the present dilemma" (2).

As Bayne-Jones put it to Weaver (1938a), Yerkes had initially been "entirely crushed by the events of the past" and had no energy or inclination to direct the laboratory's future. At this point, however, Yerkes "was no longer a crushed cur"; the Yale authorities "have been having somewhat characteristic difficulties themselves with Yerkes and that side of the problem is still unsolved."

Surely, several things were going on. Yerkes wanted to maintain an integrative focus of the work at the station and anticipated the reductionistic slide that might occur under Jacobsen. Not unimportantly, however, Jacobsen had become allied with Fulton, now Jacobsen's main supporter on the committee. Yerkes saw the possibility that *his* station would fall under the indirect control of his Yale rival. Yerkes had, indeed, not forgotten the interactions with Fulton over Tinklepaugh and probably others as well.

Fulton (1938) took a hard line; he clearly wanted Yerkes out and Jacobsen in. Although he was firm in wanting Yerkes out, he seemed not to be vindictive toward him as an individual. Indeed, it is in his February letter to Bayne-Jones that the idea of naming the station for Yerkes first finds its way into the printed record. Fulton, like the others, seemed to want to give full credit for what Yerkes had accomplished, but to want someone else to move the facility to the next level of accomplishment in changed times. Alternatively, the proposed renaming of the station may have been a bone thrown to Yerkes as a consolation prize in an effort to appear benign while seizing power.

The eventual interim solution began to emerge at this meeting. Yerkes would remain on as nominal director, but would reside full-time in New Haven and his salary would be transferred from the station. The station would be under the interim direction of someone else. The committee was split, with two members favoring the appointment of Jacobsen over Yerkes' objections and two recommending that Yerkes continue at the then current budget level.

The matter was resolved when Jacobsen (1938a) withdrew from consideration. He had wanted a stronger commitment for the future of the station than Yale was willing or able to provide. Further, he had heard about Yerkes' reactions to him. There followed an exchange of letters between Jacobsen and Yerkes in which some of their mutual misunderstandings were revealed. The essence of the disagreement was that whereas Yerkes had told Jacobsen that he would recommend neither him nor Nissen for the job, Jacobsen regarded this as different from the strength with which Yerkes had actually opposed the possibility of Jacobsen's appointment (e.g., Jacobsen 1938b).

Events were continuing to move toward a resolution. Bayne-Jones and Weaver met to discuss the situation of February 23 and the committee held its sixth meeting in New Haven. The latter meeting went smoothly, perhaps because Fulton was absent, on leave at the time. The general outline that was emerging more completely was first that Yerkes would remain for a time as the nominal director, but be gone at some point within the next five years. He would remain in New Haven and transfer responsibility for the station to a resident assistant director. The present scientific advisory board would be replaced by an administrative committee, with power to influence budgets, appointments, and uses of the station. There would be increased use of the station by outside investigators. A request would be made for a new building to serve as a physiological laboratory. Yale would request $50,000 per year and seek funds from other external sources to bring the total station budget up to the level needed to enable its increased range of activities (Bayne-Jones 1938). Bayne-Jones clearly understood Weaver in ways Yerkes did not and he was now able to effect solutions of which Yerkes, who had been able to do so earlier, was now incapable. Meeting in the absence of Fulton, the committee affirmed the basic outline discussed by Weaver and Bayne-Jones (Miles 1938b). The exchanges between Weaver and Bayne-Jones continued as the details of a mutually acceptable proposal were hammered out, point by point. At one point, Yerkes went up north from Florida to discuss the matter with Bayne-Jones in a manner Weaver was led to believe was "not particularly a jolly one" (Weaver 1938b).

On April 13, President Seymour (1938a) could finally make a formal proposal to the Rockefeller Foundation along the lines of the emerging agreement. Weaver seemed favorably disposed toward the proposal, although he requested some clarification concerning its details (Seymour 1938b). When the Rockefeller trustees finally met, however, they were unwilling to provide support at the level requested. Rather, they indicated a general willingness to fund the station with a tapered budget of $50,000 for 1939–1940, $44,000 for 1940–1941, $38,000 for 1941–1942, $32,000 for 1942–1943, and $25,000 for 1943–1944 (Weaver 1938c). The RF officials wanted the YLPB staff to seek additional funding from other sources, so that they did not have to carry the entire load. They provided $35,000 for the new building. The tone of the exchanges between Weaver and Seymour and Bayne-Jones was cooperative and friendly, with the appearance of mutual respect that never characterized Yerkes' interactions with Weaver.

The Incorporators and the Yale Corporation met and approved the revised plan (Seymour 1939). The Executive Committee of the RF met and approved the allocation of $189,000 for the tapering five-year support for the Station plus $35,000 for the erection of a physiological laboratory at the Station (Thompson 1939).

Weaver (1940) reflected: "There was a time when I feared that the best we could hope for, out of the Yale chimp project, was a retreat involving minimum loss. Through patience, firmness, extensive use of expert consultants, and through the first-rate cooperation of Dean Bayne-Jones, we managed to effect an administrative and scientific reorganization which, I believe, fully justified our further support. Even Yerkes seems satisfied!" Bayne-Jones clearly had come to the rescue as the right man at exactly the right time. For his part, with the situation resolved, Yerkes seems to have mellowed too. After a December interview with him Weaver noted that "throughout the interview [Yerkes] is more relaxed, more natural, more uninhibited, more friendly, and more cheerful than [Weaver] has ever seen him before" (Weaver 1939). Peace had been restored and the future of the YLPB seemed secure—at least for a while.

8

Finding a New Director: Karl Spencer Lashley

THE AGREEMENTS NEGOTIATED AMONG DEAN BAYNE-JONES, THE Committee on the Yale Laboratories of Primate Biology, the Rockefeller Foundation, and Robert Yerkes assured the continuation of the Orange Park facility. Two major problems lay ahead: replacing Yerkes as director and reorganizing the YLPB. The appointment of Karl Lashley would solve both problems.

Yerkes retired as Director of the YLPB in 1941 at the age of 65 and from Yale at the age of 68, the mandatory retirement age. He and his family spent most of July 1939 at the station but made only occasional visits during the next few years. As had been agreed, Henry Nissen was appointed as assistant director in 1939 and ran the Orange Park facility in Yerkes' absence.

The task of finding a new director and organizational structure fell to the Administrative Committee. The committee, chaired by Carl Hartman and including Edgar Allen, Leonard Carmichael, William H. Taliaferro, and Robert Yerkes, with the dean of the School of Medicine *ex officio,* was charged with overseeing the running of the laboratory and making recommendations to the Yale Corporation "concerning appointments, budget, policies, operation of the Laboratories, and the selection of research projects" (Administrative Committee Minutes January 29, 1939). This was sweeping power of the sort previously wielded by the director. The new dean of the School of Medicine, Francis G. Blake, was pivotal.

With Yerkes about to retire, a new organization was essential. Yale was willing to hand over control of the facility to some other institution. The committee authorized Yerkes to negotiate an agreement with the Carnegie Corporation,[1] which considered, but rejected, the possibility.

[1] It should be noted that the Carnegie Corporation of New York and the Carnegie Institution of Washington are both mentioned throughout the YLPB materials. They are not always distinguished clearly. The author has made an effort to identify the appropriate

During 1940 and 1941 the administrative committee considered Hartman, Nissen, Lashley, and Ward Halstead as possible directors. At its June 1941 meeting it finally adopted a resolution to authorize Dean Blake to invite Karl Lashley to prepare a plan and program of work at the YLPB (Administrative Committee Minutes June 8, 1941).

All indications are that Lashley was immediately agreeable. The emphasis thus shifted so that Lashley was the major actor negotiating with Yale, Harvard, the committee, and the funding organizations to develop a set of conditions under which he could direct the YLPB.

Lashley's Plan

With war raging in Europe and American entry imminent, Lashley got right to work. His preliminary statement was dated just twenty-three days after the June 8 meeting. Negotiations and revisions continued and a revised report was ready for the January 31, 1942 committee meeting.

On the surface, Lashley's plans looked a lot like Yerkes'. However, the emphasis he would place on different aspects of the overall plan would give the YLPB a very different flavor during the Lashley years from those under Yerkes.

Three changes in emphasis stand out in his July 1 draft. First, the quality of the product would be more a function of the quality of the investigators than of the biological material available. He would recruit topflight "men." Second, the range of species would be increased. Because many problems could be studied with monkeys more efficiently than chimpanzees, he would add a monkey colony. Lashley believed that chimpanzees are most useful in studies of endocrine and reproductive functions, development, dynamic aspects of behavior (i.e., motivation, emotion, and individual differences), and intelligence. Other areas of focus would be experimental and clinical neurology and developmental studies. Indeed, he would also add facilities for rodents and other "lower" animals as needed. The was necessitated by his third emphasis, to move toward a more reductionistic approach than Yerkes' integrative perspective.

A major problem for Lashley was to sell the idea to President Conant of Harvard. Emergency measures were in force, with a 10 percent budget decrease and faculty asked to move to a twelve-month commitment with-

---

referent where possible, but some ambiguity and possible errors remain. In the case of the actual allocation as Lashley assumed the directorship, it appears as though the Corporation provided funds to the CIW for support.

Photograph of Karl Spencer Lashley, the second director. Courtesy of the Special Collections Department, Smathers Library, University of Florida, Gainesville, Florida.

out additional pay. The proposal was coming at the wrong time. Conant, however, was sympathetic and agreed to support the proposal as long as some details concerning reincorporation, availability of the facility to graduate students, Lashley's availability in Cambridge for the summer terms, and the approval of the appropriate administrators could be worked out (Lashley 1941a).

By January, soon after the bombing of Pearl Harbor, the plan included a proposal that the Yale Laboratories of Primate Biology would have to be reincorporated with a new corporate board representing both Harvard and Yale and would be renamed the *Yerkes* Laboratories of Primate Biology in honor of their founder (Lashley 1942a, b).

Lashley pieced together a budget of about $65,000: $12,000 for his Harvard salary, a continuing contribution from Yale of $10,000 a year in lieu of Yerkes' salary, $25,000 from the RF, and $15,000 from the CIW. The Fels money would fill in for the first three years. Negotiations with Warren Weaver at the RF resulted in an additional grant of $118,000 for the five-year period ending June 30, 1947 (Thompson 1942). The Carnegie Corporation of New York appropriated the CIW $50,000 payable in $10,000 annual installments to the YLPB for partial support (Bush 1942). The budgetary aspects were coming together.

The primary research program would include studies of intellectual capacities, sex behavior, individual differences in temperament and emotion, the physiology of the central nervous system, and development. Lashley (1942a) diagramed his conceptualization of the interrelations among these projects.

With respect to personnel, four areas needed to be covered, each with a senior "man" and a junior investigator. Lashley would run the programs on neurological research and intellectual functions. Nissen would run the developmental program and collaborate in other projects. Young would handle research on endocrine function and sexual behavior and a fourth investigator would work on individual differences and personality.

The whole package was considered and approved at the last meeting of the administrative committee, January 31, 1942 under the committee's new chair, Leonard Carmichael. A seven-member Board of Scientific Directors (two from Yale, two from Harvard, and three from outside) replaced the administrative committee. The few matters that remained were addressed at the first meeting of the new Board of Scientific Directors (Minutes June 21, 1942). Dean Blake, the board's vice chair, George W. Corner, Derek E. Denny-Brown, Frederick L. Hisaw, William H. Taliaferro, and Yerkes were also elected. Carmichael and Yerkes were psycho-

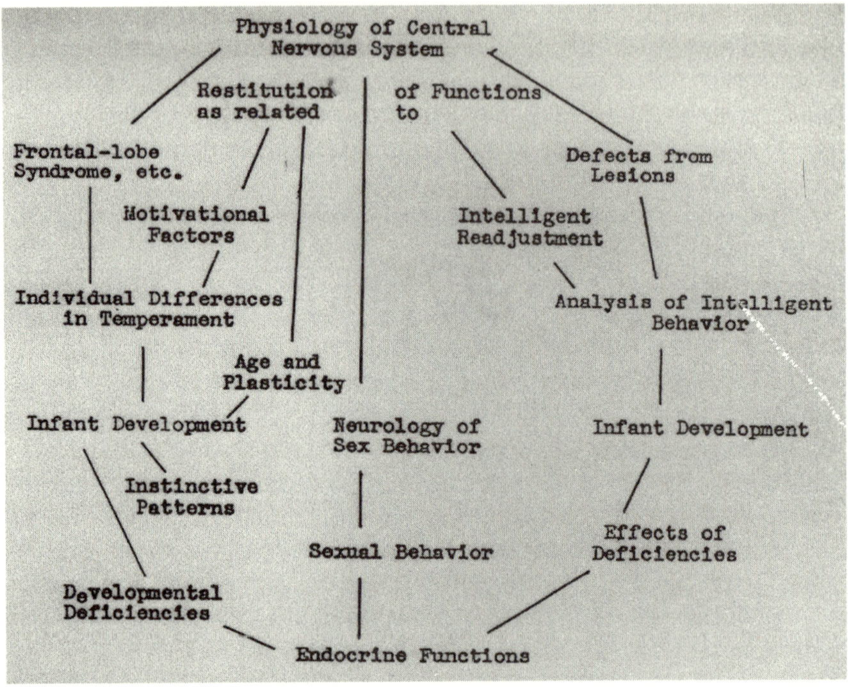

Karl Lashley's plan for a research program at the Yerkes Laboratories of Primate Biology. Courtesy of the Yerkes National Primate Research Laboratory, Emory University, Atlanta, GA.

biologists; the remaining members were of a more reductionistic bent. Taliaferro was a longtime friend of the new director, having become friendly in 1915 at Johns Hopkins.

The Corporation, formed with the duties of handling financial affairs, was distinct from the Board and included an equal number of members (two or three) from both Harvard and Yale, all university administrators. Initially, these were James B. Conant, president, William B. Claflin, Jr., treasurer, and Henry L. Shattuck, fellow of Harvard and Charles Seymour, president, Thomas W. Farnum, former associate treasurer and comptroller, and Carl A. Lohmann, secretary of Yale.

Lashley had already found his men: the scientists who would provide the core of the YLPB work in the 1940s. Young would work half-time at Orange Park. The remaining staff were individuals who would have major impact on the field: Donald O. Hebb would come from Queen's

University in Canada to work initially on personality changes after central nervous system injury but mainly on dynamic aspects of behavior. Austin Riesen from Yale would work with Nissen on the Infant Studies Program. Roger W. Sperry, a recent University of Chicago PhD and a future Nobel Prize winner, would work on nerve transplantation, initially in monkeys.

Lashley officially came on board July 1, 1942. There was a new captain at the helm. He would take the YLPB in a different direction than his predecessor.

Karl Spencer Lashley: 1890–1958

Karl Spencer Lashley was among the most influential psychologists of the century. He was a physiological and comparative psychologist of the first order. Lashley always refused to write an autobiography, although there are some unpublished autobiographical fragments (e.g., Lashley 1945a, undated). There are also several obituaries and biographical chapters (e.g., Beach, 1961; Bruce, 1991; Hebb, 1959). Some of his pivotal publications were reprinted by Beach, Hebb, Morgan, and Nissen (1960).

Lashley clearly presented himself to the world in a role of the skeptical, hard-nosed, disinterested scientist (Weidman, 1999). An historically interesting question concerns whether or not this was a genuine persona; it probably was.

Basic Life History

EARLY YEARS

Karl Lashley was born in Davis, West Virginia on June 7, 1890. As a boy, Lashley's life was dominated by books and nature. Like Yerkes, he loved his mother but disliked his father (Keefer 1958). Between 1894 and 1898 the family moved several times but resettled in Davis in 1899.

EDUCATION

Lashley attended West Virginia University, where John Black Johnston, a neurologist, taught his first course in zoology and shaped the future of the sixteen-year-old boy. In zoology he found a field in which he could combine his love of animals, freehand drawing, and his growing materialist philosophy. Lashley then worked in the zoology department under Johnston's successor, Albert M. Reese. Reese's teaching style was nondirective, pointing Lashley in a given direction and leaving him to work out the details. This would foreshadow Lashley's own nondirective style of mentoring.

After completing the AB at West Virginia in 1910, Lashley was awarded a teaching fellowship in biology at the University of Pittsburgh. He was not happy there as there was a heavy emphasis on bacteriology, so he completed his Master's degree in a year and left.

Lashley spent the summer of 1911 at the Carnegie Laboratories at the Cold Spring Harbor Biological Laboratory on Long Island. There he worked with H. S. Jennings, professor of zoology at the Johns Hopkins University. Lashley went to Hopkins, his father's choice of a school, and completed his PhD on inheritance in the asexual reproduction in *Hydra* in 1914. He found three influences at Hopkins. In addition to his work with Jennings in zoology, he had minors in psychology with John B. Watson, and in psychiatry with Adolf Meyer.

Lashley spent the summer of 1914 studying homing and instinctive behavior with Watson in the Dry Tortugas Islands off the coast of Florida. This was classic field research typical of that later conducted by the European ethologists. During his next year at Hopkins he occupied a laboratory next to W. H. Taliaferro, who would become a lifelong friend and later a member of the YLPB board. Lashley continued to work with Watson, although he quit work with the human infants Watson was studying because he "disliked the little bastards" (Lashley 1945a, 2). Lashley also came under the influence of Shepherd Ivory Franz, then at St. Elizabeths Hospital in Washington, who was studying behavioral effects of surgically created brain lesions. Much of the rest of his career can be seen as a melding of the influences of Jennings, Watson, and Franz (Bruce 1986).

PROFESSIONAL LIFE

In the fall of 1917 Lashley accepted an appointment arranged by Robert Yerkes at the University of Minnesota. He left after a year and returned to Baltimore to work with Watson for the United States Interdepartmental Hygiene Board on the effectiveness of motion picture films in educating the public concerning the dangers of venereal disease.

Lashley returned to Minnesota in 1920, protected by the new chair Richard M. Elliot from distractions that might interfere with his research.

This started a very productive six-year period in which Lashley rapidly rose to national prominence.

In 1926 Lashley left Minnesota to work at the Behavior Research Fund at the Institute for Juvenile Research in Chicago and then as a professor of psychology at the University of Chicago, remaining there until 1935.

When new Harvard University president James B. Conant sought to hire the "best psychologist in the world," Lashley was chosen. One factor in his leaving Chicago was that with the upcoming retirement of Harvey Carr as department chair he sensed pressures for him (Lashley) to become the new chair at Chicago—a position he could not tolerate as he disliked administration (C. S. Lashley 1958).

Lashley retired from the YLPB directorship at the age of sixty-five on June 30, 1955. In February 1954, while teaching at Harvard, Lashley collapsed and had to be hospitalized, diagnosed with acquired hemolytic anemia. Months of cortisone administration softened his spine. Eventually he had a splenectomy. Seemingly recovered, he and his wife traveled to Europe, where he died on August 8, 1958 in Poitiers, France.

Although he tended to belittle teaching, Lashley was a mentor for some of the leading psychobiologists of the time, including Frank A. Beach, Donald O. Hebb, David Krech, Norman R. F. Maier, T. C. Schneirla, Roger Sperry, and Calvin P. Stone (Dewsbury 2002b).

PERSONAL LIFE

In 1918 Lashley married Edith Anne Baker, a skilled musician afflicted with a severe asthmatic condition. Their only son, born in November 1919, died shortly after birth.

Beach (1961) portrayed Lashley as being uninterested and not overly able in handing finances. Nevertheless, Lashley did quite well in the stock market. It was profits from market dealings that enabled him to establish the Fund for Neurobiology to help support research in the neurobiological sciences.

Edith died in 1948. In 1957 Lashley married Claire Imredy Schiller, the widow of Hungarian psychologist Paul Schiller. Along with Claire came her two children, Peter, who was already off at college, and Christina, a source of many stories about Lashley, or "Curley" as she called him.

PERSONALITY

As described by Beach (1961) and others, Lashley was aloof without being cold. He was reserved and generally resisted intimacy with his associates. One psychologist noted that "his mind somehow seemed disembodied" (Beach 1961, 184). Although he had several very close friends, such as Taliaferro, he also had a deep sense of personal privacy.

Beach noted that Lashley had a good sense of humor and loved good plays and music. The Lashleys were fond of hosting evenings with friends playing chamber music, with Lashley playing a not-so-good cello. He even played with the Jacksonville Philharmonic Orchestra, although Elliot (1958) described him as playing "with more gusto than technique."

Lashley loved working with his hands and was a skilled carpenter and surgeon. He loved to sail and to work on his sailboat and the intricacies of sailing.

A product of rural West Virginia a century ago, Lashley held many views that are fairly judged as racist in contemporary perspective (Dewsbury 1993, 2002a; Weidman 1999). He was politically conservative. All indications are that he did not bother to vote.

As with many who held strong views about race, however, he treated those African Americans around him well—as long as they remained in "their place." Robert Pollard was an African American who worked for both Lashley and Nissen at the YLPB and in their homes. Pollard (1994) told me that Lashley "was a wonderful fellow; you know he had patience and he would sit and he would talk and explain different things to me." "He was just a wonderful man as I'm concerned."

As part of her project on the psychology of scientists, Anne Roe (1952a) interviewed Lashley, procured an autobiographical statement, and administered some projective tests (Roe 1952b). Roe confirmed the existence of Lashley's deep prejudices. She quoted him as saying "in Morgantown there were no Negroes allowed and I approved" (Roe 1952b, interview, p. 9). However, she felt this to be just a part of an overall feeling of superiority as Lashley noted that his "mother undoubtedly had the feeling that she was better than the townspeople" (Roe 1952b, interview, p. 20).

Although Lashley was also fully capable of expressing sexist views of his time (e.g., Lashley 1954a), he was progressive in his approach to women as scientists. Robert Yerkes would have no women staff members in his laboratory. Lashley added Elaine Kinder to the staff less than a year after he took over; she returned to conduct follow-up research many times.

His prejudices appear generally to have been expressed against ethnic groups but, like G. S. Hall (Diehl 1986), Lashley seemed not to hold his prejudices against individuals from those groups. In an interview Karl Pribram (1993) commented both on Lashley's reputation for anti-Semitism and the good relations between the two men, though Pribram is Jewish. Such paradoxes are not unheard of in the history of psychology.

Roe quoted Lashley as suggesting that "if you want a good description of my personality, Sheldon's description of his extreme ectomorph type [*i.e., typically tall and thin in appearance and cerebral, introverted, and asocial in orientation*] is practically exact. I read it to my wife once and her reaction was one of intense resentment. She said, 'How dared he write about you'" (Roe 1952b, interview, 3).

Roe's observations suggest a picture of Lashley as an iconoclast and a misanthrope with deep personal prejudices. Although he expressed little respect for humanity at large, in one-to-one interactions he was capable of strong positive personal relationships. He was self-centered, self-motivated, and self-contained.

Lashley valued quality—of course defined by his standards of what that meant. He was patient of those who seemed bright and sincere and intolerant of those with pretenses and little substance. He did not suffer fools gladly.

HONORS

Lashley was not fond of scientific meetings but nevertheless received many honors and was elected to various offices. He served as president of the American Psychological Association (1929), the Eastern Psychological Association (1937), and the American Society of Naturalists (1947). He was elected to the Society of Experimental Psychologists, the National Academy of Sciences (1930), the American Philosophical Society, the American Academy of Arts and Sciences, and various other organizations. He was invited to give the Hughlings Jackson lectures in Montreal and the Vanuxem Lectures at Princeton University. Lashley received the Howard Crosby Warren Medal from the Society of Experimental Psychologists in 1937, the Daniel Girard Medal of the National Academy of Sciences, and the William Baly Medal in Physiology of the Royal College of Physicians. He received honorary doctorates from the University of Pittsburgh (1936), University of Chicago (1941), Western Reserve University (1951), the University of Pennsylvania (1955), and the Johns Hopkins University (1953).

Approach and Research

CONCEPTUAL PREMISES

In contrast to Yerkes, Lashley presented himself as being the epitome of the disinterested scientist; this appears to have been an accurate portrayal. Bruce (1991), for example, quoted a letter from Lashley in which he stated that "we should not deceive ourselves with the belief that the solution of society's problems can come from science. Science can affirm means but not ends" (321).

I summarized some of the defining beliefs held broadly by Lashley and some of his prominent protegees (Dewsbury 2002b). Unlike many contemporaries, they believed that psychology was the study of both behavior and the mind. Cognitive processes were to be included in psychology. Behavioristic methods were regarded as useful, but limited and much of behaviorism was regarded as misguided (e.g., Lashley 1923). Organization, at all levels, was a key aspect of behavior and mind. Physiological analysis of the mechanisms of mind and behavior was regarded as the key to a soundly based psychology. Studies of mind and behavior were to be placed within an evolutionary framework. Research with nonhuman animals was viewed as an important part of psychology. European ethology was regarded as an important and healthy influence on psychology. The Gestalt influence was important, although there were aspects of the Gestalt approach that were deemed unacceptable. Although the development of behavior has a strong genetic influence, environmental influences interact with them and are also important. Individual differences among animals are important and should not be swept under the carpet. Although Lashley was known as a destroyer of theories, he did not oppose them categorically. He was merely a very harsh critic of the theories that had been proposed, including his own. He was an unabashed materialist.

The view that genetic factors exerted a strong influence on behavioral development is a consistent theme from his early work in genetics at Johns Hopkins into his later career. It has been suggested (Weidman 1999) that his hereditarianism drove much of his science. The argument is not convincing, however (Dewsbury 2002a).

The late nineteenth and early twentieth century saw a conceptual shift in which discussions of "instinct" went out of favor. Lashley (e.g. 1938) believed that the revolution had been misdirected. The problem with these theories, he believed, was that they hypothesized various forms of psychic energy for which there was no physiological evidence. Yet, rather than eliminating such models from their theories, psychologists threw out the notion that behavior can be innate; Lashley believed this to be misdirected.

According to some popular theories of his time, such as that of Pavlov, learning occurs by the establishment of connections among cells in the brain that were essentially the same as spinal reflexes. The system was purported to work like a telephone switchboard, with a limited number of cells making functional connections with a limited number of other cells. Lashley (1930) believed his experimental results to be incompatible with such a reflex theory. For Lashley, the central nervous system was not a mere conductor of nerve impulses but was "also an integrator of them" (Lashley 1931, 16).

RESEARCH

Lashley's range of interest, as reflected in his publication list, was extremely broad (see Beach et al., 1960). His early publications reflect the interests of his early advisors: Watson, Jennings, and Franz. Important research conducted at Minnesota and Chicago dealt with brain mechanisms of learning as, for example, in his lone monograph, *Brain Mechanisms and Intelligence* (Lashley 1929). It was this work that led to his famous principles of mass action and equipotentiality. Learning and retention by rats in mazes and related tasks was found to be affected by cortical destruction but the location of the lesion had relatively little effect (equipotentiality); rather, it was the amount of cerebral tissue that was destroyed that was critical (mass action). Note that these principles apply only to certain parts of the brain and to certain tasks.

Why Did Lashley Move To and Stay at Orange Park?

Like most decisions, Lashley's decision to move to Orange Park was multiply determined. On the surface, it is puzzling. A pure scientist, eschewing administration, accepted a job as an administrator. Further, this was the same Lashley who in 1937 had written that "from the standpoint of the comparative neurologist and psychologist, the further development of the chimpanzee as a laboratory animal is not of sufficient importance to justify the cost." Claire Lashley (1958) recalled that his attitude never changed: "the chimps themselves he found most unrewarding for research. He often said that there wasn't anything a rat couldn't do as well as one of them."

Officially, Lashley wrote to Harvard President Conant that he was "interested because it presents an unusual opportunity for an integrated study of many problems of physiology and behavior under conditions which cannot be duplicated in a northern climate" (Lashley 1941b). Surely, this was accurate; this was a research opportunity of a lifetime.

On the plus side, Lashley could handle the administrative problem by relying on Nissen to do most of the heavy jobs of running the facility, leaving himself free to do research and to provide overall direction. In addition, Lashley was not fond of classroom teaching, believing most of it to be a waste of time. The move would minimize his classroom commitments.

Lashley also had legitimate personal reasons for moving from Cambridge. Lashley (1945a) indicated that one of the reasons he moved to Cambridge was for the sailing, but it was disappointing. He found the Cambridge climate abominable (Lashley 1945a). In addition, Edith's health in Cambridge was not good; he, and others, thought that the change in climate might help her (Klüver 1942a). Lashley (1954c) wrote to Karl Pribram that "my own motives for coming here were mixed; Edith's health and my dislike of New England were certainly important factors."

Granting the sufficiency of these reasons, Lashley failed to acknowledge his problems at Harvard as a very important reason for leaving Cambridge. Harvard's courtship of Lashley was led by E. G. Boring, who was seeking a biological psychologist to help hold up the basic research approach at Harvard. Things had not gone smoothly in Cambridge. Upon arriving at Harvard, Lashley found Boring on sabbatical and appreciable administrative tasks falling upon him. Some of the promised support never materialized. In addition, Lashley went to Harvard believing that the department shared his vision for a biological psychology. In fact, although Boring shared that general view, the department was deeply divided and Lashley often found himself in the minority. The critical event concerned the tenure case of Henry Murray of the Harvard Psychological Clinic (Triplet 1992; Robinson, 1992). Murray was a personality theorist working in a psychoanalytic framework. In 1936 Boring and Lashley were strongly opposed to granting him tenure. They and others served on a committee to consider Murray's promotion and tenure. When the tenure committee was deadlocked, President Conant decided on a compromise that appeared basically in Murray's favor (Triplet 1992).

Although both Boring and Lashley opposed Murray, Lashley's intensity in doing so surprised all. Lashley went so far as to state that if Murray was not terminated, he, Lashley, would resign and return to Chicago. Conant was able to keep Lashley only by creating for him a research professorship that essentially removed him from the department. Given the situation in Cambridge, it is not really surprising that Orange Park looked appealing.

The decision to spend the rest of his career in Orange Park is a bit more puzzling. There was little, if any, beneficial effect of the Florida climate on Edith. Yet this veteran of the faculties of Minnesota, Chicago, and Harvard spent the rest of his life in what was then a backwater area removed from the centers of power in the academic world. Surely the sailing was great, and that was important. The Lashleys could play music and hear some good music in Jacksonville, but it was rarely of the world-class variety available in urban centers. However, Lashley had found a place where he was free to exercise his science as he chose with few real administrative cores. He could foster the careers of promising young scientists. As he aged and grew ill, he may have mellowed a bit and his pleasant existence appears to have been quite satisfying for him.

9

The Lashley Era in Orange Park

THE YLPB UNDER DIRECTOR LASHLEY WOULD BE MUCH DIFFERENT from the Yale Laboratories of Primate Biology under Yerkes. The changes were a function of both internal and external factors. On the outside, the scientific and funding milieu was changing and a more reductionistic approach was coming into vogue. On the inside, Lashley was a very different scientist and administrator from Yerkes. Administratively, the facility became much freer, with investigators able to pursue projects of interest to them with much less control by the Director. The utilization of the facility as only a chimpanzee research station was loosened. Significantly, invasive brain research, which had been minimal at the southern Branch under Yerkes, became an active enterprise.

ADMINISTRATION

The Board of Scientific Directors remained relatively stable during Lashley's directorship, as members were appointed and reappointed to successive five-year terms. Denny-Brown's term expired in 1945 when he was on a prolonged assignment in India. He was replaced by Wilder Penfield of the Montreal Neurological Institute. Penfield, in turn, was replaced in 1951 by J. Lawrence Pool of Columbia University. Vice-Chairman Blake died in 1952. C. N. Hugh Long came aboard in 1952. Thus by 1955, on Lashley's retirement, Carmichael was still the chairman with Corner, Hisaw, Taliaferro, and Yerkes also serving for the entirety of Lashley's directorship. Long and Pool had come aboard later and were still serving.

The corporation, made up of administrators from Harvard and Yale, and distinct from the Board of Scientific Directors, continued as the owners of the YLPB and had some control over the purse strings. It flexed its muscle, for example, in 1951, when Lashley planned three additional buildings, two of which were disapproved by the Incorporators.

With regard to administration, where Yerkes sought to control all work at the facility, Lashley preferred to delegate many administrative responsibilities, most of which went to Nissen, and concentrate on his research interests.

Lashley was a persistent and skilled critic of virtually all proposed research. At the same time, however, he encouraged its execution. According to Hebb (1980):

> As director of the laboratories, Lashley was a good man to work with. Yerkes, the first director (when the laboratories were the Yale Laboratories of Primate Biology), wanted things done in an orderly way. When I first arrived at the laboratories I found in my desk a booklet for investigators in that earlier regime that said, at one point, that originality in research is, of course important, but...and went on to list some rules and regulations concerning experiments with the animals. Lashley was at the opposite extreme. Housekeeping detail was left mostly to Nissen, and I saw no sign of any control of research by Lashley. Having assembled a staff, he left it to them to get to work; he provided what material support was possible in wartime; he argued vigorously for his own theoretical views, but he did not force them on the rest of us. (297–98)

At the same time, Lashley was also capable of a remarkable do-it-yourself approach. In an interview Karl Pribram (1993) described a situation in which he wanted to do some brain surgery on a chimpanzee and found the operating room dirty and unfit for surgery. He returned to the room to find Lashley on the floor scrubbing it. He had not called anyone but decided to do it himself. Finally, Pribram pitched in and they completed the task together.

In one memorandum, Alan Gregg of the RF summarized an interview with Lashley concerning the YLPB noting, "the interview was clear, crisp and time-saving" (Gregg 1948). This was a welcome turnaround from the ponderous meetings with Yerkes about which several RF administrators had complained. The memoranda circulated among RF officials during the Lashley administration were generally more positive, though still guarded, regarding scientific progress in Orange Park than under Yerkes.

After his February 1954 illness Lashley was greatly limited with respect to the time he could allot and be present at the YLPB. Administration became largely delegated to Nissen and Pribram.

Funding

The Lashley Years of the YLPB represent a transitional period in the history of research funding in the United States in general and the Orange

Table 9-1
Income for the Yerkes Laboratories of Primate Biology
by Directorship
(Percent)

Director(s)	Private	University	All Government	Misc.	Mean Annual Budget
Yerkes 1930–1942	79	16	0	5	$ 66,891
Lashley 1943–1955	60	23	12	5	85,352
Nissen/Peacock 1956–1959	38	0	57	5	164,754
Riopelle 1960–1962	13	0	86	1	387,745

Park facility in particular. It is instructive to examine these years in relation to the whole funding life of the YLPB. During the Yerkes years, 95 percent of the support for the laboratories came from private and university sources; there was no government funding (see Table 9-1). For the final three years for which I have data, during Arthur Riopelle's directorship, private sources contributed just 13 percent; government sources chipped in 86 percent. It was during Lashley's watch that the first Federal funds became available and this dramatic shift began. Private and university sources still contributed 83 percent of the budget but 12 percent came from the federal government. Change was in the air.

At the same time, the size of budgets grew (Table 9-1). Lashley's budgets were about $20,000 more than Yerkes'. Riopelle's budgets were five times those of Yerkes.

Another important change is that little of the money in the early years was for targeted projects. Most was for general support of the laboratory. The director was responsible for the prudent and productive expenditure of funds and had to satisfy the foundations, but he had considerable control over which projects would be conducted. This contrasts with the granting procedures that would be in force at the end of the period, when the bulk of the funds would be for targeted projects. Beginning with his 1944 annual report, Lashley differentiated unrestricted funds from restricted funds, an important differentiation.

Funding During the Lashley Years

The sources of funds during the Lashley years are presented in Table 9-2. Some of the government funding was for a World War II project to select and train oscilloscope operators that was administered through the Laboratories by Nissen. The YLPB got overhead funding. This first Federal funding was not really related to research at the YLPB.

The first government research grant to the laboratories appears in the 1948 annual report, $5,750 from the Public Health Service (USPHS) (grant MH77) for studies of different areas of the cerebral cortex involving the cutting of their interconnections—an archetypical Lashlean-type study. However, as Federal programs were just developing, Lashley would secure just $36,558 from the USPHS during his tenure as Director.

In 1952 the Office of Naval Research awarded a $33,000 contract for a study of the effects on behavior of different types of partial damage or scar formation of the cerebral cortex. Funding from the Atomic Energy Commission began in 1953.

Just as in the early years Yerkes had been the right man for Rockefeller funding, Lashley was the right man to secure funding in his time. He took the research program from the rather holistic psychobiological approach of Robert Yerkes to a series of more reductionistic, invasive studies in keeping with the times and was able to benefit as Federal funds for research were just beginning to grow.

The transition to government funding was just beginning, however. The bulk of the funding came from elsewhere. The Rockefeller Foundation contributed $30,000–50,000 to the annual budgets during this period. A total of $10,000–$12,000 was contributed by the Carnegie Institution of Washington during 1943–46 and the Carnegie Corporation during the 1947–55 fiscal years. The remainder of funds from private sources came from the Fels Fund.

Yale continued to contribute an amount equal to Yerkes' annual salary, $10,000 per year. Harvard's contribution was Lashley's salary, which remained at $12,000 with no raises during 1943–51 and leveled off at $13,000 in the year before his retirement. The pattern of expenditures did not vary greatly during this period. In fiscal year 1951, for example, over $51,000 (69 percent) of funds went to salaries; $17,931 (24 percent) went to maintenance of the facilities. Restricted funds were utilized in support of salaries and other expenses directly related to the project involved.

Table 9-2
Funding for the Yerkes Laboratories of Primate Biology During Lashley Fiscal Years, 1943–1955

Source	Amount	% of Total
Private		
Rockefeller Foundation	$ 490,000	44
Carnegie Institution of Washington	40,000	4
Carnegie Corporation	106,000	10
Fels Fund	29,339	3
Total	665,339	60
University		
Yale University	100,000	9
Harvard University	159,167	14
Total	259,167	23
All Government		
Atomic Energy Commission	28,571	3
Public Health Service	36,558	4
U. S. Navy	33,000	3
Government Overhead	29,491	3
Total	127,620	12
Other		
Committee for Research in Problems of Sex	4,700	0
Sale of Animals	4,301	0
Revolving Publication Funds	129	0
Rosalia Abreu Memorial	5,877	0
General Fund	40,099	4
Royalties from Films	69	0
Interest	2,280	0
Total	57,455	5
Grand Total	$1,109,581	100

(note: some rounding errors)

Funding Research on Irradiation

An archetypical example of the impact of funding patterns on research at the Laboratories lies in studies of whole-body irradiation, which lasted nine years beginning in 1953. The general conclusion from this research was that irradiation, especially in the higher doses, sometimes made the chimpanzees sick and interfered with motivation, but had few demonstrable effects on intellectual performance. Few of the studies conducted were ever published in full. There is little indication that the YLPB scientists had a burning interest in this problem prior to the availability of AEC funding. This availability is understandable within the context of the time, as the explosion of the first nuclear devices had a major impact on American culture (e.g., Boyer, 1985). Anxiety related to nuclear war was rampant and is somewhat difficult to understand in today's context. Nissen suggested that the interest by the Atomic Energy Commission in effects of whole-body irradiation stemmed more from the increased use of nuclear energy in power plants than in destructive explosions. To the contemporary mind viewing the program out of its proper context, however, these studies appear to have been a waste of a chimpanzee life.

Personnel

Whereas Yerkes was accused of favoring associates who were not at the top of their fields and whom he could easily dominate, Lashley tried to recruit those whom he regarded as the cream of the crop. Nevertheless, a recurring theme in Lashley's annual reports, as in Yerkes', was the difficulty of attracting the best people to the remote site in Florida.

Research Staff

With the exception of the reliable and ever-present Henry Nissen, along with William C. Young and Austin Riesen, there was a complete change of scientific personnel when Lashley assumed the directorship. Two of his first hires were among his best: Donald Hebb and Roger Sperry, both of whom had worked with Lashley at Harvard. A summary of personnel during these years is presented in Table 9-3.

By my count there were nine doctoral dissertations completed at the YLPB for various universities during the Lashley years (Birch, J. Blum, R. Blum, Chow, Gavan, McDowell, Orbach, Sperry, and Welker). Seven other staff members worked at the facility before completing the doctorate but did not do dissertations there (Dowis, Fantz, Fischer, Hutt, Premack, Shaklee, and Thompson). I cannot locate doctorates for four of the staff (C. Hayes, Jamison, Sawyer, and Wade). The remaining staff members came to the YLPB at varying times after completing the doctorate. Three each came with Chicago and Johns Hopkins degrees, and two each from Cornell, Harvard, and Northwestern. Five other universities contributed one each.

Table 9-3.
Staff Members During the Lashley Years with Birthplace, Birth Date, PhD Information, and Approximate Years of Service

Herbert G. Birch. b. April 21, 1918, New York, NY. PhD New York University, 1945. "The Relation of Experience and Motivation to Problem-Solving in Chimpanzees." MD NYU, 1960. Orange Park 1944–46.

Josephine Semmes Blum Evarts. b. November 2, 1916, Hazlehurst, MS. PhD Yale University, 1949. "Cortical Organization in Somesthesis." Orange Park 1945–50.

Robert A. Blum. PhD Yale, 1949. "The Nature of Delayed Response Deficit in Relation to the Locus and Character of Prefrontal Extirpations in Primates." Orange Park 1945–49.

Reginald B. Bromiley. b. May 22, 1911, Edmonton, Alberta, Canada. PhD Johns Hopkins, 1947. "The Development of Conditioned Motor Responses After Partial and Complete Removal of Neo-Cortex." Orange Park 1949–51 (not continuous).

Kao Liang Chow. b. April 21, 1918, Tietsin, China. PhD Harvard University, 1950. "Effects of Partial Extirpations of Posterior Association Cortex on Visually Mediated Behavior in Monkeys" Orange Park, 1947–54.

George Clark. b. May 18, 1905, Sunnyside, WA. PhD Northwestern University, 1939. "Temperature Regulation in the Cat," Orange Park, 1943–47.

James L. Dowis. b. December 3, 1924, Mankato, MN. M.S. Florida State University, 1953. PhD University of Florida, 1964. "The Effects of Evaluative and Non-Evaluative Feedback on Sequential Behavior in Psychiatric Illness." Orange Park, 1953–56 (not continuous).

Edward Vaughan Evarts. b. March 28, 1926, New York, NY. MD, Harvard University, 1948. Orange Park 1949–50. d. July 6, 1985, Bethesda, MD.

Robert Lowell Fantz. b. September 25, 1925, Muncie, IN. PhD University of Chicago, 1955. "Object Preferences and Pattern Vision in Newly Hatched Chicks." Orange Park, 1953–55.

Gloria Jane Fischer. b. April 15, 1927, Indianapolis, IN. PhD Purdue University, 1955. "The Influence of General and Specific Goal Orientations of Intentional and Latent Learning." Orange Park, 1954–55.

L(uella) Pearl Gardner. b. March 26, 1900. PhD Cornell University, 1932. "The Learning Process in Horses." Orange Park, 1945.

continued

Table 9-3. *Continued*

James A. Gavan. b. July 17, 1916, Ludington, MI. PhD (anthropology) University of Chicago, 1953. "Growth and Development of the Chimpanzee; A Longitudinal and Comparative Study." Orange Park 1950–53.

Catherine H. Hayes (Nissen) b. Milwaukee, WI. attended University of Wisconsin. Orange Park 1945–56.

Keith James Hayes. b. February 20, 1921, Milwaukee, WI. PhD Stanford University, 1947. "A Quantitative Study of the Cognitive and Emotional Effect of electroconvulsive Shock in Rats." Orange Park 1947–55.

Donald Olding Hebb. b. July 22, 1904, Chester, Nova Scotia, Canada. PhD Harvard University, 1936. "The Innate Organization of Visual Perception in the Rat" Orange Park 1942–47.

Paul J. Hutt. b. October 26, 1909, Philadelphia, PA. PhD Duke University, 1953. "Rate of Bar-Pressing by the White Rat During Periodic Reinforcement as a Function of the Quality and Quantity of the Food Reward." Orange Park 1952–54.

John H. Jamison. Orange Park 1953–54.

Murray Elias Jarvik. b. June 1, 1923, New York, NY MD 1951 and PhD 1952 University of California, Berkeley. "A Study of Some Factors Involved in Probability Discrimination." Orange Park 1951–52.

Elaine Flitner Kinder b. September 22, 1890, St. Paul, MN. PhD Johns Hopkins, 1925. "A Study of the Nest-Building Activity of the Albino Rat." Orange Park, 1943–44.

Karl Spencer Lashley. b. June 7, 1890 Davis, WV. PhD Johns Hopkins, 1914. "Inheritance in the Asexual Reproduction of Hydra." Orange Park, 1942–55.

Arnold Albert McDowell. b. May 16, 1924 Boscobel, WI. PhD University of Texas, 1958. "Transfer by Normal and Chronic Whole-Body Irradiated Monkeys of a Single Learned Discrimination Along a Peripheral Cue Gradient." Orange Park 1954–55.

Henry Wieghorst Nissen. b. February 5, 1901, Chicago, IL. PhD (Psychology) Columbia University, 1929. "The Effects of Gonadectomy, Vasotomy, and Injections of Placental and Orchic Extracts on the Sex Behavior of the White Rat." Orange Park 1939–58, New Haven 1932–39.

Walter Dorn Obrist. b. August 15, 1924, Yonkers, NY. PhD. Northwestern University, 1950. "Skin Resistance and Electroencephalographic Changes Associated with Learning." Orange Park 1952–54.

continued

Table 9-3. *Continued*

Hiroshi Odoi. PhD. University of Chicago, 1952. "An Investigation into the Functions of the Visual Cortical Areas in the Cat." Orange Park, 1953–54.

Jack Orbach. b. September 20, 1927, Montreal, Quebec, Canada. PhD Princeton University, 1955. "Effects of Removal of Visual Cortex on Learning and Retention in the Monkey." Orange Park 1952–55.

David Premack. b. October 26, 1925, Aberdeen, SD. PhD University of Minnesota, 1955. "Influence of Prior Learning of One Discrimination Component on Learning of the Second Component when the Two Components are Presented in Combination." Orange Park, 1954–55.

Karl H. Pribram. b. Vienna, Austria February 25, 1919, MD University of Chicago 1941. Orange Park volunteer.

Austin Herbert Riesen. b. July 1, 1913 Newton, KS. PhD. Yale, 1939 "Delayed Reward and Delayed Response in Chimpanzees." Orange Park, 1940–49 (not continuous)

John Shepard Robinson. b. June 25, 1920, Buffalo, NY. PhD Cornell University, 1953. "An Investigation of the Effect of the Learning of Naming Responses for Stimuli on Their Later Discrimination." Orange Park 1953–55.

George Sawyer. Orange Park 1951.

Paul Harkai Schiller. b. November 4, 1908 Budapest, Hungary. PhD Petrus Pazmany University, Budapest, 1930. "A System of Psychological Categories." Orange Park 1947–49. d. May 1, 1949, Mount Washington, New Hampshire.

Alfred Barral Shaklee. b. July 28, 1915 Chicago, IL. PhD University of Minnesota, 1943. "Psychological Effects of Vitamin B1 Supplements to a 'Normal' Diet. Orange Park 1942–43.

Roger Wolcott Sperry. b. August 20, 1913 Hartford, CT. PhD University of Chicago, 1941. "Functional Results of Crossing Nerves and Transposing Muscles in the Fore and Hind Limbs of the Rat." Orange Park 1942–46.

Robert Thompson. b. June 21, 1927, Chicago, IL. PhD University of Texas 1955. "Transposition as a Function of Stimulus Comparison in the White Rat." Orange Park 1951–52.

Marjorie Wade. Graduate student. Orange Park 1944–56 (not continuous).

Wallace I. Welker. b. December 17, 1926 Batavia, NY. PhD University of Chicago, 1954. "Play and Exploration in Chimpanzees." Orange Park 1952–54.

William Caldwell Young. September 8, 1899 Chicago, IL. Ph. D. University of Chicago, 1927. Orange Park 1939–43.

Reproductive biologist William C. Young was a critical member of the team. Together with Frank Beach, he virtually founded the entire discipline of behavioral endocrinology (Goy 1967). The only YLPB staff member to go on to earn a Nobel Prize was Roger W. Sperry, who worked in Orange Park during 1942–46 and went on to share the 1981 prize in Medicine/Physiology (Puente 1995). At Orange Park he worked primarily on the development of the wiring of the nervous system in amphibians, fish, and monkeys. Donald Olding Hebb (Beach 1987; Glickman 1996) rejoined his mentor Lashley in Orange Park in 1942. His *Organization of Behavior: A Neuropsychological Theory* (Hebb 1949b) is a classic in the history of neuroscience. Austin Riesen spent the period from 1940–49 in Orange Park except for a three-year stint in the Army Air Corps during World War II (Rumbaugh 1998). His work on the effects of environmental deprivation or enrichment in infancy on later development in primates became widely known and cited.

Paul Schiller, a native of Hungary, moved with wife Claire to the United States in 1947; after several short-term engagements, he accepted an appointment to the YLPB, where he remained until his death in 1949 (Dewsbury 1994a; Lashley 1994). Robert Thompson has been called "the true heir of Karl Lashley" (Crinella 1990; Thorne 1995). Elaine Kinder was the first woman scientist listed in the YLPB staff files. Kinder was primarily a research-oriented clinical psychologist. During her 1943–44 stay in Orange Park she initiated a long-term project on the development of chimpanzee behavior. She continued occasional work at the YLPB for some time after leaving. Keith Hayes was a Stanford PhD who joined the YLPB staff in 1947. Less information is available about wife Cathy Hayes (later Cathy Nissen). The Hayes' work with the home-reared chimpanzee Viki made her among the most publicly visible of the YLPB staff. Edward Evarts, Wallace Welker, James A. Gavan, and Jack Orbach all conducted important research during Lashley's reign.

Karl Pribram is not included in the YLPB personnel list presumably because he was a practicing neurologist and neurosurgeon in Jacksonville who volunteered time for research at the laboratories beginning in 1947 when he started spending two days a week there. He retained his association with the YLPB after moving to Connecticut in 1948. Pribram became one of the leading physiological psychologists and made many innovative contributions, both in theory and research, to the field.

Technical and Service Staff

The YLPB personnel lists include fifty-five technical and service staff who worked during the Lashley years. Perhaps the most important were the superintendents: William C. Atwater and Graham Townsend. Laboratory personnel were also fortunate to find infant caretakers who would stay for relatively long periods and provide stability in the care of the infants—a critical factor. Among the more notable secretaries were Marcia Hewey, who served from 1935–45 and Mrs. Murray Townsend, who worked from 1943–74. Mrs. Townsend, who served as a kind of executive secretary, was delegated broad responsibilities by the directors. Although they appear to have been treated well, there was, of course, a status difference between the research and support staffs that generally limited and shaped social and other contacts.

Personal Problems and Tragedies

Characteristic of the Lashley years were tragedies and marital complications. Cathy Hayes in Orange Park carried on an extensive correspondence with the Yerkes family in New Haven that is revealing even if, at times, it makes life in Orange Park sound more like a soap opera than a scientific story (Dewsbury 1994b). In 1949 Paul Schiller took a brief leave from the YLPB to go to Harvard to work with B. F. Skinner. Always the adventurous type, he went on a skiing trip to Mount Washington, New Hampshire. As Lashley put it, "an earlier thaw had iced a portion of the trail and in his rapid descent he slipped to his death over the precipitous West Wall. So ended prematurely a scientific career of great promise" (Lashley 1994, 313). Eight years later Lashley and widow Claire Schiller would marry and enjoy a brief, but happy, marriage.

A second loss of life occurred soon thereafter when Robert A. Blum was drowned at Vilano Beach near Orange Park on July 1, 1949. Blum had been part of a party of eight scientists who went to the beach for a picnic and swim. Several got in trouble in the ocean currents. Although Keith Hayes was able to rescue Blum from the ocean, he could not be revived.

There were illnesses and injuries as well. Nissen had two serious operations in 1947. Lashley was taken ill with acquired hemolytic anemia. His role at the YLPB was greatly reduced for the remainder of his term. During much of the 1950s Nissen, like Lashley a heavy smoker, suffered from emphysema. Robert Yerkes suffered a heart attack in April 1954 and died in February 1956.

The Orange Park environment was not benign to marital relationships. As noted by Orbach (1982, 23), "in this setting, marital relationships were frequently strained, sometimes broken, and otherwise rearranged." The small group of academicians was in close daily contact and isolated from others of similar backgrounds. In March 1954 Cathy Hayes, for example, wrote to Yerkes of the decision by Keith and her to divorce after 12½ years of marriage (C. Hayes 1954a). The next month she would announce to Yerkes that she was the new Mrs. Henry Nissen (C. Hayes 1954b).

Animals

The animal population changed under Lashley as he built up colonies of several other species of nonhuman primates. Larger numbers of animals could be purchased, used in studies of brain function, and sacrificed for histological analysis. Further, Lashley supported research with animals other than primates as well. These included rats, cats, rabbits, and amphibians.

With his interest in studies of brain function, it was practical for Lashley to use monkeys rather than chimpanzees. Almost immediately after assuming the directorship, Lashley established a colony of spider monkeys. In 1947 he purchased twenty-four rhesus macaques and six mangabeys for studies of cortical function. In subsequent years, he purchased assorted other monkeys.

The success of the chimpanzee breeding program ebbed and flowed year by year, as had been the case under Yerkes. When Lashley assumed the directorship in 1943 the laboratories owned forty-nine chimpanzees, one of which was on loan elsewhere. At the time of Lashley's retirement the total was sixty-six chimpanzees, fifty-nine of which were on-site.

Like the animal numbers, the overall health of the colony ebbed and flowed. In 1945 Lashley reported a problem with bacillary dysentery and other diseases; sulfa drugs brought the problem under control temporarily. Dysentery, however, was endemic to the colony in 1945. Lashley noted that this posed a risk to the staff as well as the chimpanzees; in 1946 three people came down with the amebic dysentery. The next year Lashley could report that the health was improving. By 1948 the problem had again become sufficiently serious so that an epidemiologist, Dr. William Burrows of the University of Chicago, was brought in to conduct a study of bacillary dysentery. In 1951 the Director reported that overall health was good and that dysentery was under control.

Chimcrackers were still in use but there were many food supplements: "Every day there were several feeding rounds made of the compound by different workers. Bananas were given in the morning, bread, apples, orange slices, grapes, sweet potatoes and carrots were handed out to each animal during routine feedings. A dipper full of milk was given to each of

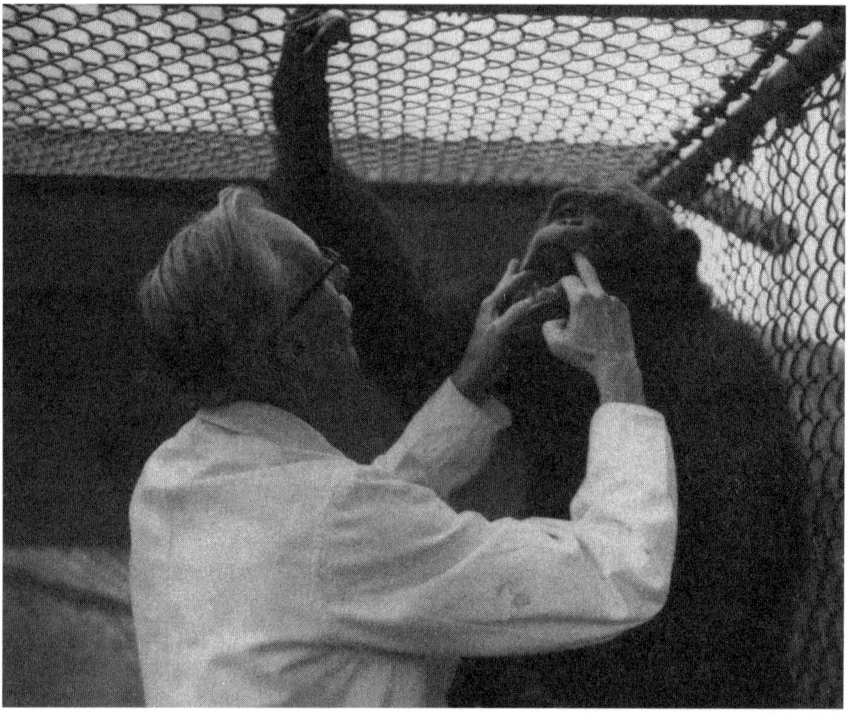

Henry W. Nissen inspecting Bokar's teeth. Courtesy of Catherine Nissen.

the younger animals in the outside compound. The infants in the upstairs nursery in the main building were fed bottled milk" (Welker 1996).

Welker also described a procedure of "cabbaging," where Nissen would make the rounds, feeding each chimpanzee a wedge of cabbage. With this practice, Nissen could conduct a routine daily examination, even training the animals to make certain postural adjustments so that he could examine body parts. Typical of a gifted animal worker, Nissen talked to the animals, called each by name, and developed a unique rapport with them.

There were several proposals to convert the facility into more of a breeding and supply station than the research facility that it was. For example, in 1945 a proposal was turned down to serve as a receiving and distributing center for rhesus macaques.

Physical Plant

The physical plant continued to grow. By 1953, Lashley could inventory twelve buildings in addition to the garage, pump house, and various storage sheds: the Administration building, Laboratory Building, Shop and Kitchen, Quarters Building, M Building, N Building, K Building, E Building, Hospital Building, Monkey House, Superintendent's House, and a Small Bungalow. The lettered buildings were for the housing and testing of chimpanzees of varying ages.

The addition of a spider monkey colony in July 1942 necessitated the addition of a facility to house them. Construction was completed in August of a frame building with two laboratory rooms, outdoor cages, and heated shelters capable of housing between fifty and seventy-five monkeys. The cost was under $3,000.

The next year plans were approved for the construction of a heated concrete shelter for the chimpanzees. During wartime, however, workmen for the construction of the facility could not be found. An addition to the infant quarters, an experimental shed, and additional housing for monkeys were built, largely by members of the staff. A small quarters building for young animals was built in 1951.

An annual set of problems concerned the maintenance, repair, and renovation of the already existent facilities, which suffered in the Florida climate.

The Impact of World War II

The YLPB, like the rest of the country, was affected by the breakout of world war. It affected housing, staff, and the kind of work done at the facility. The YLPB was situated between the Jacksonville Naval Air Station, just northeast of Orange Park, and Camp Blanding to the southwest. Improved roads led through Orange Park and made it a desirable place for military personnel to live. The influx of personnel reduced the available housing for scientists.

Several key members of both the scientific and support staffs left for military service. The war made them more difficult to replace than previously.

The YLPB also was involved in several projects of direct military relevance. Perhaps the most interesting was Nissen's project with oscilloscope operators. "Some 15 psychologists were employed on the project and 64 separate reports of research were made. The project involved expenditures in excess of $172,000" (Lashley, Annual Report, 4). Nissen's role appears to have been primarily fiscal oversight; Donald Lindsley handled most of the hands-on research. The connection to the YLPB was minimal. However one of the participants, Robert S. Daniel (1994) recalled, "The most interesting consequence of such a curiosity came from a visitor to our lab who stared into each room and finally asked 'Where are the monkeys?' Mystified, one of us said 'What monkeys?' then he said 'I thought you were training monkeys to operate radars.' "

In another project, Sperry was assigned to work with Paul Weiss on studies of nerve regeneration (Contract OEMcmr-221). Weiss spent about three weeks working in Orange Park. In his Annual Report of 1953, Lashley noted that "important results were obtained in studies of recovery of function after suturing and interchanging severed nerves . . ., and the work contributed to the development of a tantalum sleeve as a substitute for pieces of artery in the suturing of cut nerves" (5) (e.g., Sperry 1945a). Lashley himself served on a committee formed by Vannevar Bush charged with the task of developing mechanical aids for the blind and partially sighted. The committee worked primarily on the production of guidance and reading devices and to improve lenses.

Life in Orange Park

Lashley's directorship of about thirteen years occurred during momentous times. As he took office the United States was at war; it was only months after the Pearl Harbor attack. Franklin Delano Roosevelt was the United States president and the Germans reached Stalingrad. "White Christmas," "The White Cliffs of Dover," and "Praise the Lord and Pass the Ammunition" were the songs of the day. Lashley's tour as director lasted through the conclusion of World War II, the Korean War, the Cold War, and the election to the presidency of both Harry S. Truman and Dwight D. Eisenhower. Development of the atomic bomb, the McCarthy anticommunist hearings, and Bobby Thomson's home run all occurred. The NSF and NIMH came into prominence in funding science. The world had changed. Surely, these and other major events affected life in Orange Park. In some respects, however, the town seemed unchanged, still isolated from world events.

As was true during each of the directorships, the scientists varied greatly in their reactions to life in Orange Park. There were many complaints of the summer heat in pre-air-conditioning Florida, of the insects, and of the isolation from academic life. Orbach (1982) and some others "found the laboratories a lonely and isolated place where there was nothing to do but work. North Florida was a cultural wasteland: and Orange Park itself was in the middle of nowhere." (23).

Young's biographer described his subject's experience at Orange Park: "The isolation, the slower pace of the work, the closeness of interpersonal involvements, and the nearly stereotyped rivalries that encumber any small community depressed him and slowed progress even further. When he left in 1943, he carried with him an abiding fear of isolation from the professional setting and a distrust for enterprises not closely integrated with the academic community" (Goy 1967, 6).

The perennial problem of finding adequate housing, exacerbated with the development of the nearby military installations, was a recurring difficulty. Lashley chose to live on the outskirts of Jacksonville, at "Fiddler's Cove."

Others, however, adjusted well to the environment and looked back with great pleasure on their experiences there. There were pleasant staff lunches each day, many gatherings and outings for laboratory staff, sailing, and musical adventures. A cluster of lawn chairs had been placed in front of the main building and provided the location for daily group lunches filled with lively discussion. Badminton courts were installed and this became a popular pastime for laboratory personnel. The many visi-

Lunch in Orange Park. This gathering was typical of the informal atmosphere that prevailed at the Laboratories. Included are Catherine Hayes Nissen and Henry Nissen (far right). Courtesy of Catherine Nissen.

The Hebb family in Orange Park: an illustration of family life during the 1940s. Courtesy of Mrs. Rene Wilkie and Mary Ellen Hebb.

tors, from students to Alfred Kinsey, Konrad Lorenz, and John Dewey, provided welcome interactions.

Helen Riesen (1994) had many pleasant recollections. Like Lashley, the Riesens were sailors; they owned a 17-foot sailboat and sailed on Doctors Lake and the St. Johns River with colleagues, visitors, and friends. Also like the Lashleys, they were musical, Helen at the piano and Austin on clarinet. Helen Riesen accompanied vocal students and taught piano to numerous children, including Ellen and Jane Hebb and Dora Nissen. They had many friends from both the laboratories and the community.

Occasionally, issues of race surfaced. Jane Hebb recalled her parents' struggle with the inequalities afforded blacks in Orange Park. Robert Pollard (1994), an African American who worked at the YLPB, recalled the staff members as being generally benign but termed many of the locals "die-hards." Pollard classified Superintendent Atwater as being in that group.

Bill Foley (1988) wrote of some of the community perceptions, as viewed from the local Boy Scout camp, of the YLPB during the early 1950s: "It was a solitary place, behind a high wall topped by electrified wire. It was shrouded in mystery compounded by myth. . . . The first thing every kid heard the first night he got to Camp Echockotee was that an ape had escaped from the Monkey Farm which is what kids called the place. . . . Forget science. Forget Yale. Forget intelligence tests and frontal lobotomies. We are talking gorillas here. Great apes. Mad, hairy monsters, thrashing through the underbrush, smashing all before them, their only intent to crush the spleens of fleeing children" (Foley 1988, F-1, F-4).

Filling Lashley's Void

During much of the time of Lashley's illness, beginning on February 7, 1954, Nissen and Pribram functioned as acting directors. However, Lashley remained, at least nominally, as the Director until his retirement on June 30, 1955. By that time, Pribram had withdrawn his candidacy to become Lashley's successor. Although Pribram (1993, 1994) indicated that he had been a codirector, it appears that the duo functioned only as acting directors and that Nissen became the third director on July 1, 1955.

Several documents support the view that there were indeed serious negotiations about a codirectorship but that they never came to fruition (Lashley 1954b, d). At the Board of Directors meeting of March 15–16, 1954, for example, it was agreed that Nissen and Pribram would function as codirectors upon Lashley's retirement and would develop a research plan to be presented to funding agencies to secure a firm financial basis for the continuation of the YLPB. Meanwhile, the Yale administration had become disenchanted with the idea of supporting a laboratory in Florida that, with Yerkes' retirement, bore little relationship to activities in Connecticut. At a May 20 meeting of the Corporation the matter was considered in relation to a program prepared by Nissen and Pribram. The decision was made to continue the then existent system and defer further considerations of change. "This motion also included the approval of the appointment of Doctors Nissen and Pribram as Administrative and Executive Directors of the Laboratories respectively" (Pool 1954). A key aspect was the assurance that the goals of the YLPB were to advance understanding of human psychology and psychiatric problems.

Pribram (1954b) was upset by Pool's summary, which appeared to restrict the kinds of research that could be done at the YLPB based on the decision of the Corporation, rather than scientific personnel. In his letter to Lashley he noted that "I will not accept a position at the Laboratories under duress" and that "there is no need for you to consider me in regard to the appointment unless you can persuade me that this atmosphere of suspicion and threat is an erroneous one."

The matter grew more complicated. Soon Pribram had soured on the idea of a codirectorship (Pribram 1955). Nissen had altered one of Pribram's research proposals without consulting him. Pribram saw the prospect of becoming a full-time administrator, a possibility that held little appeal to the research-oriented neurosurgeon. Both Harvard and Yale clearly wanted to divorce themselves from the laboratories, leaving the future of the facility in doubt. The whole situation was complicated throughout with the need of Pribram to maintain his staff position in Connecticut in order to avoid the military draft. The YLPB Board wanted Pribram to spend significant time in Orange Park.

By the May 8 meeting of the Board of Directors, "Drs. Pool and Long reported that Dr. Karl Pribram was no longer a candidate for the codirectorship of the Laboratories." The Board then "Voted: to recommend to the Corporation the appointment of Henry W. Nissen as Director of the Laboratories for the year 1955–56 at an annual salary of $12,000".

10
Research Under Lashley

As MIGHT BE EXPECTED, THE RESEARCH TOPICS EMPHASIZED UNDER Lashley were quite different from those under Yerkes. An estimate of the approximate number of publications during the two administrations devoted to different categories is presented in Table 10-1. There was, of course, considerably more research on neural mechanisms. The emphasis on sexual behavior and physiology, higher cognitive processes, social behavior, and sensory function was reduced compared to the previous administration.

Studies of Development

Studies of normal development were begun in 1939 when Nissen secured money from the Fels Fund for his study that continued through the life of the YLPB and beyond. Nissen's study of behavioral development was designed to run parallel to the studies of Arnold Gesell at Yale on humans. First results were reported in the book by Riesen and Kinder (1952). The chimpanzees were found to have more rapid motor development than humans, but fall progressively behind in perceptual and manipulative abilities. Some of these differences appeared secondary to morphological differences; others appeared to be strictly behavioral (e.g., chimpanzees tended to explore the environment with their lips and tongue rather than with their hands).

Physical anthropologist James Gavan spearheaded studies of physical growth (e.g., Gavan 1953). Gavan found that chimpanzees require less time to complete their growth than do humans and that females require less time than males. The chimpanzee gestation period and childhood are longer relative to humans, and infancy and adolescence are relatively shorter. Birth order has no direct effect on birth weight, although birth weight decreases with maternal age (Gavan 1952).

Table 10-1
Differences in Emphasis (% of Total) in Research Publications During the Yerkes and Lashley Administrations

Topic	Yerkes	Lashley
N	180	192
Nutrition, Metabolism, & General Physiology	6.1	3.6
Pathology: Neural and General	1.1	1.0
Psychopathology	3.3	0.05
Sensory and Perceptual	10.0	2.6
Development	8.9	10.9
Neural Mechanisms	7.8	34.5
Sexual Behavior and Physiology	16.1	6.8
Social Behavior	11.7	2.6
Learning	13.3	12.0
Motivation	1.7	3.6
Higher Cognitive Processes	11.1	4.2
Miscellaneous	8.9	17.2

Note: Each article is counted once only. Thus, a study of brain function and learning would be included under "brain function." These data should be interpreted with the realization that some areas may seem underrepresented. Note some rounding error.

Rearing a Chimpanzee in a Human Environment

Associated with efforts to determine the pace of normal development were studies in which manipulations of rearing environment were analyzed. These involved both enrichment and deprivation. The objective was to help separate genetic from environmental effects.

Surely the most famous of the studies during this period was that of Keith and Cathy Hayes, rearing chimpanzee Viki in their Orange Park home. The study followed up that of Kellogg and Kellogg (1933) and the never-published study of Glen Finch. The couple took the chimpanzee on September 1, 1947, three days after her birth and raised her, as far as possible, as a human child. They had hoped to study Viki through to sexual maturity, but Viki died of encephalitis on May 11, 1954.

Cathy Hayes's (1951) *The Ape in Our House* was written for a general audience with a rather informal style that concentrated on the Hayes' many experiences with Viki; the more probing studies were published in journal articles. The most complete summary of tests of Viki's cognitive capacity was not published until seventeen years after her death (Hayes and C. Nissen 1971). Viki was highly extroverted, loved both manipula-

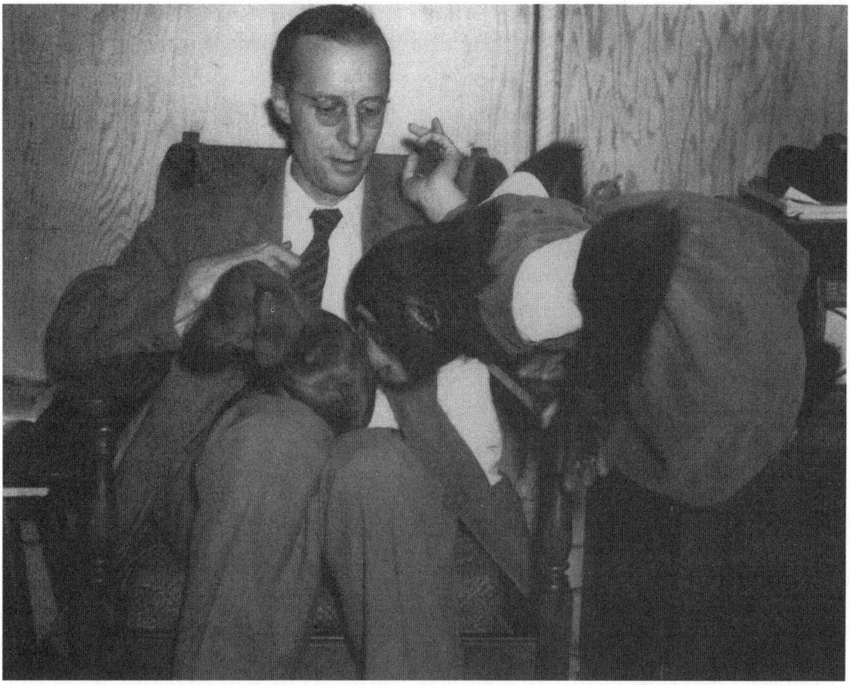

Chimpanzee Viki with Reginald B. Bromiley and his dog. Courtesy of Catherine Nissen.

tive and social play, and was much less vocal than are human children (Hayes and Hayes 1951).

The Hayes were especially interested in studying Viki's cognitive and language capacity. Formal training began with string-pulling problems when Viki was twenty-one weeks of age. The Hayes' guarded conclusion was that Viki's intellectual development was about the same as that in children for the first three years, as long as tests of verbal ability were excluded. They suggested that chimpanzees have evolved a sophisticated capacity for culture that must have been favored in the natural environment (Hayes and Hayes 1954).

A speech-training program was begun when Viki was five months old. She was not an adept learner; "it took her five months to learn to produce a hoarse, staccato grunt, quite unlike her normal spontaneous sounds" (Hayes and Hayes 1951, 107). By shaping her lips as she vocalized, they were able to get her to learn several words, like "mama," "papa," and "up." She could later make these without having her lips shaped for her. They were unable to teach Viki much more than that and concluded that chimpanzees lack the brain mechanisms necessary for human speech. Of course, more recent studies have shown chimpanzees capable of rather sophisticated language-related behavior, but not via vocalization and not as complex as that of humans.

An interesting footnote emerged in my interview with Karl Pribram (1993). He told me that when Viki was born they had tried to take her from her mother, Vera, who smashed Viki against a wall. Pribram and Lashley decided that they did not want this animal, who probably suffered brain damage, for their studies of brain function. When the Hayes sought an animal for their home-rearing study, Lashley, who was skeptical of the value of the study, was happy to give them Viki.

Effects of Sensory Deprivation

The basic procedure in this set of studies was to raise chimpanzees in such a way as to restrict their use of one of their senses. The objectives were to try to separate genetic from environmental influences and to determine the effect of deprivations on the development of later capacities.

Austin Riesen spearheaded a considerable program of rearing animals with altered visual input (Riesen 1950). He found that rearing in darkness for 16 months did not affect basic visual reflexes but little else was normal. Riesen found that it was not the absence of light per se, but the absence of *patterned* light that was essential in the deficit. Among those following up this research was Fantz (1965).

In a study of restricted tactile input, one chimpanzee was deprived of a variety of stimuli by raising it with its arms and legs enclosed in cardboard tubes from the age of four weeks to thirty-one months (Nissen, Chow, and Semmes 1951). The restricted animal showed marked deficits in tactual-motor coordination and the location of tactile stimuli

These studies, and various others, represent an integrated, though necessarily limited, program of research on the normal development of behavior and morphology and of the effects of both enrichment and restriction. Not surprisingly, effects of both nature and nurture were suggested.

Sex Physiology and Behavior

Work on reproductive physiology and behavior followed up the studies of the Yerkes years. Virtually all of this work was done in the early years of the Lashley administration. Once Young and Clark left, they were not replaced with workers on reproductive behavior or physiology and the program lay dormant.

Young continued his program of systematic work on the relationship between the hormones and both reproductive physiology and behavior. In a classic study of behavioral changes in female chimpanzees through their menstrual cycles and after ovariectomy, Young and Orbison (1944) concluded that chimpanzees lie somewhere between monkeys and humans with respect to the extent of effects of cyclical variations on behavior. The results contradicted those of some earlier studies. Other research was directed at the physiology of reproduction (e.g., Clark and Birch 1948).

In a series of studies in which they followed up some of the earlier work of Robert Yerkes and his associates, Clark and Birch studied hormonal effects on social, rather than sexual, behavior (Birch and Clark 1946, 1950; Clark and Birch 1945). They analyzed the roles of androgens and estrogens in dominance in both male and female chimpanzees. Birch and Clark (1950) suggested that they could explain the effects of swelling on dominance "without resorting to such *ad hoc* assumptions as the privilege granting concept of Yerkes" (190). In his 1953 Annual Report, Lashley called this series of studies "the first clear demonstration of the immediate effects of internal secretions on the non-sexual behavior of any primate" (18).

Because of Nissen's difficulties in writing up his research, the task of reporting results of tests from the growth study of sexual behavior in chimpanzees was left to Riesen writing much later (1971). Riesen concluded that a social learning process is essential for the full expression of mating behavior in chimpanzees. The period between 3 months and 2½ years appeared critical. Mother-rearing and peer-group interactions facilitate explorations and sociosexual behavior.

Instinctive Behavior of the Chimpanzee

Lashley found the category of "instinctive behavior" to be useful in summarizing some of the research at the YLPB. Schiller (1951) studied the drawings of Alpha, the first chimpanzee born at the YLPB. He found that Alpha was fond of scribbling in patterns similar to those of human infants aged 12–18 months. When presented with stimuli with some printed figures on them, she tended to fill in blank spaces. Schiller, with a Gestalt psychology background, interpreted this as directed at closure of the total configuration of the elements.

Hebb's name is generally associated with cognitive, rather than affective, aspects of behavior although he was interested in a wide range of problems. In Orange Park, he concentrated on affect. Hebb (1946a) noted that there were stable individual differences among the YLPB chimpanzees that greatly resemble human temperament. Hebb referred to *emotions,* by which he meant essentially deviations from a typical individual's baseline behavior. He suggested that we use the same processes in understanding animals as in recognizing emotions in other humans and in ourselves. Hebb was apologetic about the subjective nature of his hypothesis and wanted to avoid implications of conscious intent but believed that these stable patterns were real.

Hebb (1946b) analyzed fear in chimpanzees by observing their reactions to a variety of objects, including an adult chimpanzee head, a chimpanzee cadaver, a lifelike human head, and a replica of a coiled 24-inch snake. He thought that with fear, "an essential feature of the stimulating conditions is the divergence of the object avoided from a familiar group of objects, while still having enough of their properties to fall within the same class" (268). The chimpanzee or human must learn what is familiar and the normal range of variation therein. He suggested that fear entails a disruption of temporally and spatially organized cerebral activities" (274).

Analysis of Personality

Hebb's research on temperament, or personality, in chimpanzees was stimulated by similar studies in humans. Hebb (1949a) studied the reactions of different individual animals to familiar caretakers, a disguised person manifesting great timidity in approaching the animal, the same person in a different disguise acting aggressively toward the animal, and various inanimate objects. He found some consistent patterns in the responses of individuals and regarded these as "of the same kind as the differences of human temperament" (203). In a sense, he was merely quantifying what had already been known by workers at the YLPB.

Tough-minded Karl Lashley cast a somewhat different light upon the results. In the 1953 Annual Report he concluded that "the plan of investigation, which is also that of almost all studies of human personality by testing methods, was faulty; most of the behavior measured by the tests proved to be variable from day to day and no cross-section of activities gives a reliable picture of the behavior to be expected over a period of months or years" (21). One must wonder why there were two such different spins on this research. It is possible that conflict over Hebb's (1949b) classic *Organization of Behavior,* which appeared at this time, might have been an underlying source of irritation (Orbach 1998). In brief, Hebb had invited Lashley to be a coauthor of the book but Lashley declined, as he disagreed with some of Hebb's positions.

Experimental Studies of Learning

Studies of learning were more sparse during the Lashley years than previously. A representative study was that of Hayes, Thompson, and Hayes (1953), who found chimpanzee performance in learning sets, or learning to learn, not to be dramatically different from that in either rhesus monkeys or humans. Lashley and Wade (1946) reported data that they used to attack one of Lashley's favorite whipping boys—the neo-Pavlovian theory of stimulus generalization during learning.

Nissen, Blum, and Blum (1948) reported results of a study with a procedure that later became quite popular—matching to sample. In essence, the chimpanzee had to choose between two stimuli, one of which was a replica of a third stimulus present at the time. The animals showed positive transfer of training from the initial stimuli to new pairs. Nissen

Hiroshi Odoi testing Verb, Viki's older brother in a conditional matching problem.

was unable to distinguish among three possible explanations for the result.

Paul Schiller conducted several comparative studies of detour learning. In a study conducted at nearby Marineland of Florida (Schiller 1949), for example, he found that octopuses were capable of withstanding delays of about one minute in delayed-detour learning as long as they were able to maintain their postural orientation during the delay.

Evolution of Intelligence

In an effort at comparative assessment of intelligence, Gardner and Nissen (1948) studied the performance of chimpanzees in a task previously used by Gardner with a variety of other species. It entailed first associating a signal with approach, then making various adjustments as the situation changed. Initial learning was better in such species as horses, cows, and sheep than in chimpanzees. In the later transfer tests, however, the chimpanzees outperformed "human aments (both imbeciles and idiots)" (163). [It must be remembered that at that time such terms were used as legitimate category descriptions.] Gardner and Nissen found no large differences among the groups and suggested that "there is no regular increase in the rate of simple habit formation at progressively 'higher' levels of the mammalian series" (163).

Birch (1945a) sought to understand the role of motivational factors in affecting performance by young chimpanzees in the classic stick and box-stacking problems made famous by Köhler. He concluded that intermediate motivational levels were most effective. In an often-cited article, Birch (1945b) considered the effect of prior experience on performance in stick problems and concluded that the solution depended in large part upon the previous experience of the animal. Solution entailed "the integration into new patterns of activity of previously existent part-processes developed in the course of the animals' earlier activities" and that "any interpretation of insight, in situational terms alone, or even predominately in situational terms, is invalid" (382).

Schiller (1952, 1957) used similar problems a few years after Birch. Schiller's chimpanzee animals would play with sticks and join them spontaneously as part of their play. When the same individuals were required to join the sticks to reach an incentive, they were unable to do so, suggesting that behavior from the context of play did not transfer well to the context of work. Schiller believed that chimpanzees possess a certain repertoire of innate manipulative patterns. "Their *application* to external stimulus configurations is something that must be learned" (Schiller 1952, 178).

Although he was primarily interested in the neural correlates of behavioral evolution, Lashley (1949) addressed issues regarding the evolution of intelligence in an article on "Persistent Problems in the Evolution of Mind." He wrote that "evolutionary changes between marsupials and man have been almost exclusively an increase in the capacity to discover significant relations among the elements of a situation and to adjust behavior in terms of a maximum number of such relations" (41). Nissen published a series of articles providing perspective on these issues at various times. In a wide-ranging and important chapter in the *Handbook of Experimental Psychology,* Nissen (1951) considered the axes for behavioral comparisons in comparative psychology. He suggested, in a similar vein, that "it is in the cognitive rather than in the motivational aspects of behavior that we find the significant axes of behavioral evolution" (380). More specifically, he added that "Behavioral evolution is characterized by increasing complexity. With ascent in the phyletic scale, the effective environment expands spatially and temporally and provides an ever-wider range of behavior determinants. This expansion follows upon the organism's ability to integrate and organize. It is only with the advent of 'culture' that the effective environment itself expands rapidly from generation to generation" (Nissen 1951, 381). What is especially notable is the fact that this was written at a time when behavioristic approaches appeared dominant in psychology and before the so-called cognitive revolution.

Neurological Investigations

Clearly, it was studies of brain function that were dearest to Lashley's heart. Chimpanzees, however, were not ideal subjects for such work. When the work was first started, there was no available atlas of the chimpanzee brain to provide the needed anatomical basis. Lashley's disappointment with the YLPB studies of intelligence and personality left him believing that the appropriate behavioral measures were not in place either. The neurological studies were quite detailed. Therefore, I shall not try to convey them in all of their subtlety, but rather to convey the general nature of the research.

Lashley was especially interested in learning and other higher processes and these functions appear concentrated in the outer mantle, or cerebral cortex, of the brain. A number of cortical regions have direct connections from various sensory systems or to various motor regions, and thus are termed sensory or motor areas. Between these sensory and motor areas lies the so-called associative cortex, which had been thought to be the locus of memories and higher processes. It was these association areas that most fascinated Lashley and where the YLPB efforts were concentrated.

However, the program as a whole was more comprehensive than this and was a reasonably well integrated attack on the anatomy and function of virtually all of the cerebral cortex in monkeys. In the process, a reevaluation of the principles of mass action and equipotentiality, developed by Lashley in his work with maze learning in rats, was necessitated. It must be remembered that much of the work relied upon lesion techniques wherein selected tissue is removed and the function of that tissue is inferred from the resulting deficits. That was the preferred method of the day. The method is imperfect with many possible sources of deficiency other than that from the loss of the specific tissue (e.g., Orbach 1959). Results must be interpreted in that context.

Lashley's Emerging Views

In summarizing the emerging general picture, Lashley dismissed theories of control based on reflexes or conditioned reflexes in favor of more dynamic mechanisms. The brain is intrinsically active and stimuli must have their effects in the context of that ongoing activity. Complicated functions, such as abstraction and generalization, can be carried out by the sensory areas and appear inherent in the structure of nerve networks. Lashley laid out his more complete theory in several general publications (e.g., Lashley 1950b, 1951, 1958d, 1998).

In one of his best-known papers, that on the serial-order effect (Lashley 1951), he argued that, for example, the skilled movements of a concert pianist could not be controlled peripherally. There simply was not time for these repetitive interactions between the brain and the fingers to act. Rather, a central pattern generator in the brain generates these sequences of behavior. Bruce (1994) showed how this was a long-standing position of Lashley that can be traced as far back as 1917. Lashley's was an important paper in the field of motor control and it also played a supporting role in the development of cognitive science.

Beach (1961) wrote that "a major source of motivation which influenced Lashley throughout most of his professional life was the desire to formulate a comprehensive theory of learning" (180). But it was the failures of the theories that struck Lashley. In one of the more oft-quoted Lashley passages he reflected on his long search for the locus of the memory trace, or engram. "I sometimes feel, in reviewing the evidence on the localization of the memory trace, that the necessary conclusion is that learning just is not possible" (Lashley 1950a, 477–78).

Lashley's final views were presented in his Vanuxem Lectures at Princeton University in 1952 and in a lecture at the University of Rochester in 1957. Both have been preserved in the archives of the University of Florida and the former has been published by Orbach (1998), along with his controversial views regarding the interactions between Hebb and Lashley concerning the nature of neural activity. Lashley viewed the brain, and particularly the cerebral cortex, as a continuous network of cells so complexly interconnected that, via direct or indirect routes, any cell is capable of exciting any other. Behavior is determined by the actions of thousands or millions of cells not necessarily located near to each other and is a function of the overall pattern of firing, rather than of the action of small, localized centers. The networks of cells are diffuse. Lashley liked to compare this action to the waves on a lake: "Such an active system can be conceived as somewhat similar to the interplay of waves on the surface of a lake, with a sequence of waves caused by the wind, on which are smaller ripples from gusts, traversed now and again by waves from passing boats. The wave systems do not destroy one another but summate so that the effective intensity at any point is determined by the combination of crests and troughs which are passing at the moment" (Lashley 1998, 291). Such an analogy is of mixed value. It conveys Lashley's view that the summation of activity from successive waves is critical to function. However, the mechanical nature of the analogy reminds one of Köhler's field theory of spreading electrical current. The difference is that Köhler referred to a mechanical wave of electricity moving across the cortex without regard to specific circuits. Lashley did much to discredit this model. Rather, Lashley's view was that the activity took place in many directions along established neural pathways. It differed from other views in that Lashley viewed the systems of cells affecting any one activity as being very widely distributed in the brain and reduplicated therein.

The cerebral cortex, for Lashley, was "a thin network of nerve cells divided into a number of sensory areas with an intervening expanse of unspecialized nervous tissue separating them" (308). Excitation of a sensory area spreads to the associative network, which is undifferentiated, except for some bias in the areas adjacent to those sensory areas. There is a reduplication of activity throughout the wide-ranging field. That is why small lesions rarely have major effects on memory; memory is multiply and diffusely represented in the cortex.

Anatomical Studies

It was necessary to complete basic studies of anatomy before other work could be properly interpreted. Lashley and Clark (1946) paved the way for the research with an anatomical study of the cerebral cortex of spider monkeys. Whereas some anatomists had divided the cortex into over 100 regions, Lashley and Clark found great individual differences and suggested that the number of functional regions is much smaller.

The group also was interested in the connections between lower structures and the cortex. Chow (1950) showed that, contrary to dogma, there are extensive organized connections between a part of the thalamus, the pulvinar, and posterior associative cortex. Similar studies revealed other pathways between the lower structures and association cortex (e.g., Pribram, Chow, and Semmes 1953).

The staff also followed up a pet project of Lashley, the counting of cells. Chow (1951a), for example, provided estimates of the number of cells in the auditory systems of monkeys. Lashley used such counts to suggest that all cells in the brain must be continuously active and that learning must entail some change in the pattern of firing.

Studies of the Function of Association Cortex

THE TEMPORAL LOBES

In a number of studies it was shown that visual discrimination tasks learned before operations are lost after temporal lobe lesions (e.g., Chow 1951b). Chow and associates worked to understand the exact nature of the deficits. They found, consistent with other results, that the monkeys could reacquire the habits at a normal rate (Chow 1952). Although some of the additional results are difficult to interpret, they suggest that additional training facilitates the recovery of habits previously learned. Chow and Orbach (1957) used four visual tasks with stimuli presented only very briefly in a tachistoscope. No deficits were found. These data suggested that lesioned animals do not have problems with all visual tasks; rather, the difficulty is a function of the manner in which the stimuli are presented.

PREFRONTAL CORTEX

The prefrontal cortex was of special interest because of its function in humans and the earlier work of Carlyle Jacobsen that had shown that monkeys with prefrontal lesions were specifically deficient in the delayed response task. Wade (1952) observed similar effects. Robert Blum (1952) dissected this effect further and found that it was the region of the frontal lobes that is connected with the dorsomedial nucleus of the thalamus that produced most of the effect. Orbach and Fischer (1959) found that operated monkeys could be trained to perform sufficiently on the delayed-response task. They suggested that the monkeys do this with various mnemonic devices, such as positioning, and that distraction interferes with their performance in a manner not characteristic of the intact animal. They pointed to situational factors that affect the disturbances: immediate memory impairment, an exaggerated tendency to respond to spatial cues, and chronic vasodilation.

THE ASSOCIATION AREAS

Lashley summarized the emerging view of the functions of association cortex. He believed that it is a location for memory storage and that the association cortex is not necessary for perception and simple generalization. As best he could characterize the emerging picture, "The tasks failed in the absence of the association areas are difficult to characterize but roughly they may be said to involve the perception of relations among relations. An illustration may serve to clarify this statement: The animal without posterior association cortex can quickly learn to choose a red food box and avoid a green; similarly to choose between black and white objects. He will fail, however, when the task is complicated and he is required to choose the red box when both boxes are on a white tray and the green when both are on a black tray" (Lashley, Annual Report, 1953, 34). Specialization appeared minimal with regions near to sensory areas being somewhat more involved in functions related to those senses than others.

THE INTERACTION OF CEREBRAL AREAS

Lashley believed it important not only to study the functions of the various areas, but their interactions as well. This could be done by interrupting the pathways connecting them. Wade (1952) found that cutting the connections between the frontal lobes and other cortical areas had little effect, whereas cutting their connections with lower centers produced severe deficits. These and other data led Lashley to conclude that the main functional communication among cortical areas was not across the cortex itself, but via more indirect routes through subcortical structures. Although the exact pathways were not clear, he believed that many routes were available for such communication.

Sensory Cortex

Lashley was less interested in the functions of the sensory cortex itself. Evarts (1952a, b) found that removal of the auditory cortex did not interfere with the discrimination of tones with different pitches but did produce deficits in more complex tasks. Although a pre-operatively established discrimination was lost, the animals could be retrained to discriminate successfully, although not at the level of intact animals.

Orbach (1959) reviewed work on the striate, or visual, cortex at the back of the brain. Monkeys with striate lesions showed pronounced deficits in visual function but could be retrained to discriminate in a light-dark test. Thus, the region is not necessary for all visual discrimination. In addition, some nonspecific effects of the lesions were observed, particularly with respect to nonvisual orientation when moving through a maze. Other, seemingly related, performance was unaffected. Orbach suggested that there are some functions that are general and can be affected by lesions almost anywhere in the cortex and others that are more specific and affected by more circumscribed lesions.

Motor Cortex

Part of the cerebral cortex is clearly involved in the control of movements. The issue of interest to Clark and his associates (Clark and Ward 1949) concerned whether this area was organized in terms of individual muscles or whole movement patterns. By implanting electrodes in the appropriate regions and stimulating conscious monkeys they found evidence in favor of the hypothesis that it is movements, not muscles, that are represented at this level.

Theories Based on Topological Projections

Lashley had little use for the theory of Gestalt psychologist Wolfgang Köhler, according to which the cortex functions in perception via electrical fields of current flowing across the cortical surface. Several approaches were used to disrupt any such fields in order to show that specific neural connections, rather than a general field of excitation, was critical. Lashley, Chow, and Semmes (1951) inserted electrical conductors, such as gold foil, into the visual areas of monkeys to disrupt current flow and found no effects on vision. In other studies, currents were passed through the frontal lobes via implanted electrodes during delayed response performance without effect. Lashley developed his own theory about radiating and reduplicated interference patterns (e.g., Lashley 1998).

Growth and Specificity of Nerve Fibers

The work of future Nobel Prize-winner Roger Sperry was among the most important conducted at Orange Park during the Lashley years. There are direct connections between the sensory organs, such as the eyes, and receptive areas in the brain. Sperry was interested in how these systems get wired properly through development. His basic method was to interrupt the afferent nerve fibers by cutting them and forcing them to grow through a scar so that there would be a random assortment of central connections. Sperry chose amphibians on which to experiment because of their regenerative capacity. He found that even after he cut the connections between the eye and the appropriate brain areas, the regenerating fibers linked to the proper areas (e.g., Sperry 1947). This was true even when Sperry rotated the eye 180 degrees in the orbit at the time that the optic nerve was cut. He got similar results with the vestibular nerve and organs of balance (Sperry 1945b). The results suggest that there is considerable chemical specificity as the nerve fibers grow toward the receptive areas. Lashley (1958e) would later write that "the best work done during my tenure was Sperry's study of eye transplants in amphibia."

Investigations of Primarily Medical Interest

One of the contributions of the YLPB to medicine was to provide occasional animals for medical research at other sites. Most proposals, such as efforts to infest the entire colony with polio or mumps, were rejected as impractical. The work of Dr. William Burrows on immunity to bacillary dysentery was discussed in chapter 9.

Research Bearing on Psychiatry

A joint project was developed between the YLPB and the Rockland State Hospital in Orangeburg, New York to conduct research bearing on psychiatry. Nissen took a leave of absence from Orange Park to serve as research director at the hospital. Unfortunately, the program had to be abandoned after eight months because of civil service regulations in New York State.

Hebb (1947) considered spontaneous neuroses in chimpanzees. He presented a wide-ranging discussion of neuroses in humans and nonhuman animals and developed a theory for them. Hebb discussed two case histories of behavior he regarded as neurotic in Orange Park chimpanzees. The first concerned Alpha who, Hebb noted, was generally well adjusted emotionally. However, he wrote that "The exceptions were Freudian: a definite avoidance of any carrot that was forked or double-rooted, and of long, narrow cylindrical pieces of food such as okra; and a marked tendency to masturbate against the water faucet of her cage, whenever human beings came in sight" (3). Later, Alpha developed a pattern in which she appeared afraid of large pieces of food. This then developed into an alternating pattern of fearing the food and fearing the one who fed her (Hebb) and who had never harmed her. Kambi showed a pattern of great mood swings between periods of euphoria and profound depression lasting months at a time. Hebb proposed that both fear and neuroses are the result of disruption of the functional organization of the cortex resulting from conflict, sensory deficit, or metabolic disturbance.

Other Projects

Various other projects merit mention. During these years interest in exploration and play had grown greatly in psychology. The psychology of the time was dominated by learning theory and learning theory had been dominated by attention to basic need-drive systems (e.g., hunger, thirst, sex drive). Nissen (1954a) presented a wide-ranging discussion of the drive concept.

The realization that some behavior occurred in situations unrelated to the primary drives forced a reconsideration of some of the principles that had been developed. Welker conducted his dissertation work for the University of Chicago on play and exploration in chimpanzees in Orange Park (Welker 1956a, b). Novelty was the major factor producing interest. Interest in objects declined as a session progressed in a kind of satiation. The animals also preferred large objects, those that moved or were movable, and complexity.

With support from the United States Atomic Energy Commission, researchers at the YLPB studied the effects of whole-body irradiation on behavior. The early results suggested that the irradiation at the levels used produced very little effect. During the Lashley years, the program was under development and it was continued into the next administration at the YLPB. Studies of blood group factors (e.g., Wiener, Gavan, and Gordon 1953, 1958) were also conducted.

Marineland of Florida, the first of the sea aquarium facilities, is located relatively near to Orange Park. Hebb worked with the Marineland staff on the behavior of captive bottle-nosed dolphins (McBride and Hebb 1948). The published article includes important early observations on vision, hearing, vocalization, sleep, dominance relations, sexual behavior, behavior at parturition, play, and fears. The authors concluded that the emotional and motivational behavior of dolphins lies somewhere between those of dogs and chimpanzees. They were struck with the quickness with which newborn dolphins are able to swim to the surface and breathe. The authors were skeptical as to whether dolphins should be treated as a "higher" species.

Overview and Summary

In 1950 Lashley provided George W. Gray of the Rockefeller Foundation an overview of the accomplishments of the Orange Park facility (Lashley 1950c). He began by noting the importance of first showing that chimpanzees could be maintained and bred in captivity and of a survey of their sensory and motor capacities, nutrition, growth and reproduction, diseases, mental capacities, and social behavior. Lashley wrote that "more is now known concerning the chimpanzee than of any other animal, excepting rat and man" (1). He then outlined six "problems that have been studied advantageously" (1) in Orange Park. The first was the evolution of behavior and included a wide range of studies such as those of instinctive behavior, social behavior, play, and problem solving. Second was sex physiology and behavior, where he highlighted the some of the differences in chimpanzees from so-called lower mammals. The third area concerned "emotionality, temperament, and motivation" and dealt primarily with the research of Hebb and Kinder. Fourth was "the structure of intelligence" including factor analyses, comparative studies, and studies of effects of brain injury. The work of Herbert Birch and Paul Schiller was highlighted. The fifth area concerned cerebral physiology, with an emphasis on the interaction of different functional areas. Not surprisingly, Lashley elaborated a good bit on progress in this area. The sixth and final area was problems of medicine that included work on infantile paralysis and bacillary dysentery.

Lashley concluded by evaluating the anthropoid program. He sounded familiar themes. Because of their expense and the time required to study them, research on chimpanzees was justified only when it could not be done with other species. He admitted that chimpanzees may be uniquely valuable for work on sex physiology, comparative studies of intelligence, and cerebral function and perhaps in the fields of nutrition, aging, personality, and psychopathology. However, Lashley clearly believed that much of the work could have been done with monkeys although he noted that "this was not evident when the work was undertaken" (7). He wrote that "if I were asked to name the most important single contribution from the Laboratories, I should select Jacobsen's discovery of the reduction of temper tantrums in the chimpanzee by brain lesions" (7). Lashley cited the work of Freeman and Moniz as showing how such work could be applied.

Lashley began his last paragraph with a sentence that may hold for the entire history of the YLPB and for much basic research in general: "The great value of the work here is not in spectacular discoveries but in the steady accumulation of data on systematic problems" (8).

11

After Lashley: The Nissen-Peacock Years

THE FINAL DECADE OF THE ORANGE PARK FACILITY WOULD SEE three directors and an acting director. There would be a major change in sponsorship and, finally, the laboratories would be moved to Atlanta.

As Henry Nissen finally assumed the directorship in 1955, the funding of research, cultural climate, and the town of Orange Park were all changing. He would have to face the withdrawal of support from the historical sources, the retreat of Yale and Harvard from sponsorship, finding a new sponsor, personal problems including divorce, deteriorating health, difficulties in recruiting top staff members, and some internal strife. It would be difficult to fill the shoes of a Karl Lashley. Nissen managed to keep the YLPB afloat but perhaps at a very great cost to himself. The deaths of the first three directors, occurred within the space of a little over two years within this turbulent time in the facility's history. Nissen was replaced by acting director Lelon Peacock. During his term as director, Nissen would move the YLPB research program away from Lashley's emphasis and back closer to its historical roots.

THE CONTEXT

The "fabulous fifties" are remembered as a time of conformity and materialistic priorities. This was the era of "Ozzie and Harriet" and "Lucy and Desi." The fifties were turbulent as well. Joseph McCarthy held his hearings into alleged communist influences, Julius and Ethel Rosenberg were executed, and President Truman fired General Douglas McArthur. The civil rights struggle came into prominence with the Little Rock school desegregation controversy, and the *Brown vs. Board of Education* decision. Rosa Parks sparked the Montgomery, Alabama bus protests and Martin Luther King, Jr., emerged as a civil rights leader. The Russians

Photograph of Henry W. Nissen, the third director, in the shadow of Robert Yerkes. Courtesy of Dora Nissen Mann.

launched the first Sputnik satellites, the Korean War was fought, the American suburbs spread, and censorship battles were fought over D. H. Lawrence's *Lady Chatterley's Lover.* The fifties saw the arrival of Miltown, Alfred C. Kinsey's study of sexual behavior in human females, *West Side Story, My Fair Lady,* Marilyn Monroe, *Playboy,* Elvis Presley, the Cha Cha Cha, and the "beatnik" movement. The first hydrogen bomb was exploded, further forcing the threat of nuclear destruction into the collective mind set. This was also the time of the wiring of the nation for television, the occurrence of the "baby boom," and the rise of Fidel Castro (Halberstam 1993).

Closer to home, the town of Orange Park was growing. What had been a town of about 660 people when Robert Yerkes founded the YLPB in 1930 nearly quadrupled to over 2,600 by the end of the 1950s. Holly Point, fronting on both the St. Johns River and Doctors Lake, was developed. Several other developers followed. Blakey (1976) regarded Orange Park as already having become a bedroom community for Jacksonville by the early 1950s.

Directors

While Karl Lashley was ill and approaching retirement age the operation of the YLPB was in the hands of Henry Nissen and Karl Pribram. An increasingly central activity of the Board of Scientific Directors, as Lashley's June 30, 1955 retirement date approached, was the search for his replacement.

The New Director: Henry W. Nissen

With Pribram's withdrawal from a possible co-directorship, Nissen became the leading candidate. However, the Board was not fully convinced that Nissen was the man for the job as director. No one questioned his scientific acumen or his knowledge and skill in working with chimpanzees. What concerned the Board were his rather quiet nature and the perception that he would not be forceful enough outside of the facility in seeking funds and attracting new staff. Where Yerkes and Lashley were bold, Nissen was quiet and careful. Nissen (1954a) wrote that "it pays to be patient, to be sure one's I-beams and rivets are sound before building a skyscraper" (282). Nissen was both cautious and precise even down to the metaphor he selected.

Nevertheless, Nissen appeared to be the obvious choice and was named the new director as the Board voted "to recommend to the Corporation the appointment of Henry W. Nissen as Director of the Laboratories for the year 1955–56 at an annual salary of $12,000" (Board of Directors Minutes May 8, 1955).

Henry Wieghorst Nissen was born February 5, 1901 in Chicago, Illinois. Young Heinz, as he was called, attended LaSalle-Peru Township High School in LaSalle, Illinois (Carmichael 1965; Dewsbury 1998a; Koehler 1960). After a year working as a bookkeeper and newspaper reporter, Nissen attended the University of Illinois in Champaign-Urbana, majoring in English, and received his BA degree in 1923. With the death of his father in 1920, the family finances were strained and Nissen took time off before attending graduate school, working as a sales correspondent and statistician during 1923–26.

While working in New York, Nissen enrolled in night school at Columbia University and there met comparative psychologist Carl J. Warden, under whom Nissen received his PhD. Yerkes invited him to join his research program upon hearing him make a meeting presentation.

Nissen had been at the YLPB since 1939 and was perhaps the most important of all staff members to the efficient operation of the facility. He was generally acknowledged to know more about chimpanzees than anyone else and to have an unusual rapport in working with them. He is rarely remembered today but, in spite of his reticence, was regarded as sufficiently important by his colleagues to merit election to the prestigious National Academy of Sciences of the United States.

His first marriage, to Jane Marian Stowlby, ended in divorce. He had two daughters by the first marriage. In 1954, not long before his appointment as director, Nissen married the former Cathy Hayes, the second marriage for each. Although all appears to have been quite blissful in the marriage for a while, it eventually broke down. On August 12, 1956 Nissen wrote to Robert and Ada Yerkes that Cathy had resigned from the laboratory and left. He added that the initiative had been hers (Nissen 1956). The next year he added "I understand that Cathy had a baby girl on May 17" (Nissen 1957a). The issue of paternity was not addressed. John Falk (1994) described the tense social situation as "a very heavy Tennessee Williams kind of dynamic going on at the lab."

In the meantime, Nissen was having serious health problems. Nissen was a long-time, heavy smoker; his emphysema often left him gasping for breath. He was hospitalized in 1957. Nissen wrote to Ada Yerkes "my dyspnea—resulting from emphysema—became quite a nuisance this spring and I spent two weeks in the hospital soon after the Board meeting" (Nissen 1957a). He was hospitalized again with bad stomach lesions the next year.

Nissen's health problems were compounded by an accident at the YLPB. A standard procedure at the facility was to remove newborns from their mothers so that they could be reared in the nursery. In order to do this safely, the mother had to be sedated. In July 1957 chimpanzee Helene, described as a good-natured animal, gave birth and Nissen went to remove the baby. As Nissen entered the cage too soon after administration of the tranquilizer when the animal was still slightly groggy, she reached up, grabbed Nissen's leg, and slammed him onto the concrete floor causing damage to his back (Falk 1994; Peacock 1993; Rogers 1994). This was not an outright attack, as the animal did not follow up its action. As might be expected, this compounded Nissen's health problems.

Nissen took his own life with a drug overdose on April 27, 1958. Lashley (1958a) wrote: "He had been ill for a long time; not fully recovered from the spinal injury of last July, severe emphysema and recently hospitalized with ulcers and a suspicion of cancer. He had had bad luck in choice of a staff and was breaking in an entirely new group this winter—seriously overworked. He told me, three weeks ago, that he felt that he could not go on as director and I suggested ways of reducing his load. Last Sunday and Monday he did not go to the Labs. [*sic*] and did not answer the phone. On Monday afternoon the Townsends managed to break into his house and found him in bed."

Lashley called Nissen "one of the most lovable persons that I have ever known. Too kindly, too modest, too sensitive to make a success of life . . . he felt the blows too keenly" (Lashley 1958a).

Clearly, there were many sources of stress in his life. Charles Rogers (1994) told me that he thought Nissen had just reached the end of his rope. He would break out in hives every time he had to write a grant proposal or when the benign Board of Scientific Directors was to visit. Rogers added that Nissen was too late for his time. He reported data in the slow and casual manner of fifty years before him and was unhappy in the emerging publish-or-perish environment.

Acting Director Lelon J. Peacock

At the meeting of the Board of Scientific Directors held less than two weeks after Nissen's death, the Board "VOTED to recommend to the Corporation the appointment of Dr. Lelon J. Peacock as Acting Director of the Yerkes Laboratories of Primate Biology, Inc." (Board minutes May 10–11, 1958).

Lelon J. Peacock was born in Brevard, North Carolina on May 25, 1928. He attended Berea College and the University of Kentucky, receiving a PhD in psychology from the latter in 1956. He served in the U.S.

Photograph of Lelon J. Peacock, interim director. Photograph by the author.

Army Medical Research Laboratory during 1954–56 before going to Orange Park beginning November 1, 1956.. Thus, he was appointed acting director at the age of 29, less than two years after completing the PhD and moving to Orange Park. He moved to the psychology department of the University of Georgia in 1959 and retired as a Professor Emeritus in 1990.

Robert M. Yerkes

As long as he was alive, Robert Yerkes remained interested in, and concerned about, the laboratories. Cathy Hayes Nissen remained a faithful correspondent and kept him up to date on life in Orange Park. Yerkes also corresponded with Carmichael, chair of the Board of Scientific Directors and often offered his advice.

Robert M. Yerkes died in New Haven on February 3, 1956. Ada Yerkes' (1956) telegram to Carmichael the next day read simply: "ROBERT LEFT US LAST NIGHT AFTER ONE DAY IN HOSPITAL HE HAD BEEN WEAKENING RAPIDLY FOR THE LAST FEW WEEKS WE ARE NOT HAVING ANY PUBLIC SERVICE."

Karl S. Lashley

After Nissen assumed the directorship, Lashley remained in Jacksonville and stopped in at the YLPB only as his moods and health dictated. In 1957 he married Claire Schiller; the marriage appears to have brought them both great joy. Lashley (1958a) wrote to Ada Yerkes that he and Claire had had an adventurous trip to California, Vancouver, and Alaska the previous summer and had come home via Vermont. He added that in the summer of 1958 they would go to Portugal to visit Claire's parents. He would never return from the trip; Lashley died August 8, 1958 in Poitiers, France.

Personnel

Scientific Staff

Many of Lashley's staff members had departed before Nissen assumed the directorship on July 1, 1955. Several of the remaining staff members remained for brief periods that reached into Nissen's tenure. The scientific staff members during this period are listed in Table 11-1.

Table 11-1
Staff Members During the Nissen/Peacock Years with Birthplace, Birth Date, PhD Information, and Approximate Years of Service[1]

Janeth Turner Carpenter. b. December 1, 1926. PhD Florida State University, 1959. "Generalization in Verbal Conditioning." Orange Park 1956.

Paul B. Carpenter. b. February 24, 1928. PhD Florida State University, 1959. "The Effects of Sensory Deprivation on Behavior in the White Rat." Orange Park, 1956.

Richard K. Davenport, Jr. b. Charlotte, NC. PhD University of Kentucky, 1956. "An Investigation of the Drinking Behavior of the White Rat." Orange Park 1957–65.

John L. Falk. b. December 27, 1927, Toronto, Ontario, Canada. PhD University of Illinois, 1956. "The Effect of Hypertonic Stomach Loads Upon Fluid Intake in the What Rat." Orange Park 1956–57.

Charles B. Ferster. b. November 1, 1922. Freehold, NJ. PhD Columbia University, 1953. "The Effect on Extinction Responding of Stimuli Continuously Present During Conditioning." Orange Park 1955–57.

Roger T. Kelleher. b. December 28, 1926, New Haven, CT. PhD New York University, 1955. "Discrimination Learning as a Function of Reversal and Nonreversal Shifts." Orange Park 1955–57.

Emil W. Menzel, Jr. b. April 16, 1929 Central Province, India. PhD Vanderbilt University, 1958. "Variability of Behavior in Problems of Different Types and Different Levels of Difficulty." Orange Park 1957–65.

Lelon J. Peacock. b. May 25, 1928, Brevard, NC. PhD University of Kentucky, 1956. "Brain Lesions and Conditioning in the Rat." Orange Park 1956–59.

Charles M. Rogers. b. April 27, 1923, New Haven, CT. PhD Yale University, 1954. "Hypothalamic Mediation of Sex Behavior in the Male Rat." Orange Park 1956–65.

John S. Stamm. b. November 29, 1919. PhD University of Southern California, 1950. "The Relationship Between Reaction Times to Light and Electroencephalographic latency Times of Blocking of the Occipetal Alpha Rhythm." Orange Park 1955–56.

Nissen made three appointments in 1955. The first two would set the tone for much that happened in the laboratories during Nissen's time as director. Some context is required to understand these. Recall that Nissen and Pribram functioned as acting co-directors for a period. Although B. F. Skinner had been active in psychology for a number of years, he was gaining increased attention during the 1950s. Pribram was attracted to

Skinner's approach and the possibility of using his techniques of operant conditioning in his research. He had an extensive correspondence with one of Skinner's close associates, Charles Ferster, and floated the idea of hiring Ferster to Nissen (Pribram 1954a). Nissen responded favorably, noting that he liked the idea of studying "the effects of very specific and controlled experiences on later behaviors" and "the idea of developing practicable automatic training devices" (Nissen 1954a). Nissen added a caveat that was somewhat prescient: "My impression is that some people have the greatest admiration for the Skinner approach but that many more are highly critical and even contemptuous of it. The latter view might be accentuated when considered in connection with, or applied to, such a 'highly organized' animal as the chimpanzee" (Nissen 1954b).

The details were worked out and Ferster began his official tenure at the YLPB on July 15, 1955. He left for Indiana University in August 1957. Keller wrote that "As a person, Charles embodied many contradictions. He was an unobtrusive man who could make things happen; a shy man who could speak his mind; a kindly man who could be hard; a sober man with a twinkling eye and a bag of little quips. A friend described him as 'the most skillful-insightful awkward man I have ever known'" (Keller 1981, 301).

A second scientist who would work in the Skinnerian tradition, and actually arrived a month before Ferster, was Roger T. Kelleher. Nissen's third 1955 appointee was John S. Stamm, who also became attracted to Skinner's science. Yet another future proponent of operant methodology once in Orange Park joined the staff the next year. John L. Falk knew little of operant psychology, having been influenced by Hebb at McGill and attracted to the YLPB because of the studies of early experience being conducted there. He became fascinated by the hard-nosed approach to science of Ferster and Kelleher and the possibilities for using the techniques in research with drugs (Falk 1994). The three scientists who came to identify with the operant approach were gone by the end of the summer of 1957.

The operant conditioners were balanced by a quartet of scientists who worked more in the tradition of Yerkes and Nissen and who came to form a more lasting core for the staff of the Yerkes Laboratories. As noted above, Peacock joined the staff in November 1956. Charles M. Rogers arrived two weeks later. He arrived from Yale, where he had worked with Frank Beach. Rogers remained with the Yerkes Laboratories through the move to Atlanta before leaving for the University of Guelph in 1972. Through his sixteen-year tenure, Rogers became a mainstay of the research program, working especially, but not exclusively, in the developmental research program.

Richard K. Davenport joined the staff the next year and also worked primarily in the developmental program and moved with the laboratories to Atlanta. Later, like Nissen, he took his own life.

Another stalwart of the program was Emil W. Menzel, Jr., who arrived in August 1957. Menzel remained associated with the YLPB until the move to Atlanta, at which time he joined former Director Arthur Riopelle at the Delta Regional Primate Research Center in Louisiana. Menzel became well known for his research on learning and cognition and for his insightful articles relating to primate behavior.

Technical and Service Staff

The YLPB personnel lists include thirty-five technical and service staff who worked during the Nissen-Peacock period. As always, the various laborers, technicians, cooks, secretaries, and caretakers were critical to the functioning of the laboratories.

Graham Townsend continued as the superintendent and his wife Murray Townsend as office manager. There was some feeling among the research staff that Murray Townsend gained increasing power in the running of the laboratories with the turmoil in the director's position and that she controlled resources that might otherwise be thought to be under the control of the directors. Whereas Lashley (1958e) was critical of the scientific staff, he regarded the service staff as "excellent." He wrote of the Townsends that "both are gentle, unassuming people who have made a career of the Labs. and are completely dependable."

Other key support staff included Annie Jones, who served as the caretaker in the nursery from 1945–65 and Arthur Chester, who was a technical assistant during 1955–64.

The board did what it could to improve working conditions at the facility. For example, beginning in 1956 it grappled with problems of health insurance. This, and many other such problems, was typical of the kind of detailed administration necessary to run a facility such as this.

Board of Scientific Directors

Leonard Carmichael continued to chair the board and was quite active in negotiating arrangements for the continuation and development of the laboratories. Corner, Hisaw, Long, and Taliaferro continued throughout the period under consideration. J. Lawrence Pool resigned on May 10,

1958. Yerkes served until his death in 1956; Lashley was elected to replace him on March 10, 1956 and served until his death. In 1957 the board was expanded with the addition of Howard M. Phillips and Arthur P. Richardson of Emory University; Phillips stayed for just a year, after which he was replaced by Geoffrey H. Bourne, future director of the YLPB. The board was returned to seven members with the departures of Pool and Lashley.

Transfer to Emory University

One of the major issues to be dealt with during this period was that of the academic affiliation for the laboratories. This issue interacted with several others. For example, funding from the NSF was contingent upon affiliation with an appropriate institution. It will be recalled that with Yerkes and Lashley departing, both Yale and Harvard saw decreased relevance for their programs of the remote facility in Florida and considered terminating support. As Lashley (1958e) would put it: "Angell was succeeded by an historian, Yerkes antagonized the faculty and Yale lost interest in the Labs." On February 28, 1952 Yale President A. Whitney Griswold wrote to Harvard President James B. Conant "to inform you officially of our decision to terminate Yale's contribution to the annual budget of the Yerkes Laboratories of Primate Biology" (Griswold 1952).

Although the annual contributions of $10,000 ceased effective July 1 of that year, Yale and Harvard still shared joint custodianship of the facility. There was some feeling in New Haven that his too should be terminated. At its March 1954 meeting the board noted that Lashley had been approached by the newly founded Health Center of the University of Florida about a possible affiliation but, as the Health Center was just evolving and not in a position to advance concrete research proposals, the possibility was unattractive. Brandeis University received brief consideration. Carmichael volunteered to explore the possibility of affiliation with the Rockefeller Institute; Corner would explore a possible affiliation with the Carnegie Institution. Pool suggested the possibility of Emory University or with a group of southern universities.

The next step is difficult to document. Because Pool had the closest contact with the Yale administrators, was the only member of both the Board and the Corporation, and had been the one to suggest Emory to the Board, it is likely that he advanced the possibility of a transfer to Emory University to Yale officials. During a period of confidential exploration

Long, of Yale, was the only Board member whom the Emory officials took in their confidence (Phillips 1955a). Negotiations took place among President Goodrich C. White of Emory University, President A. Whitney Griswold of Yale University, H. M. Phillips, Dean of the Emory University Graduate School, and others. Phillips visited the laboratories in April 1955 and Yale University in June of that year and prepared a report on his findings (Phillips 1955b). The Emory faculty and administration were enthusiastic about the possible affiliation. The main concern was over possible financial commitments that might be incurred by Emory (Phillips 1955a). Nissen assured Phillips of his willingness to work in affiliation with Emory (Phillips 1955a). Yerkes (1955a) replied that he was open-minded, as he saw both advantages and disadvantages to affiliation with Emory.

Phillips' report included a summary of the administrative structure of the YLPB, the physical plant and equipment, a general summary of the research program, and a financial statement. Phillips concluded with a list of five advantages to Emory (the international acclaim of the laboratories, the unique animal colony, the fit with then-existing Emory programs, enrichment of the graduate school, and the additional interest in Emory that could be generated) and three disadvantages (the remote location, financial obligations, and the need to increase the graduate school budget in order to operate the facility).

At its meeting of March 10, 1956, attended by Phillips, the Board of Scientific Directors passed the following by unanimous vote: "RESOLVED, that the Board of Scientific Directors expresses its wholehearted approval of the transfer of the Laboratories from Yale University to Emory University, in a manner that will be satisfactory in all respects to both institutions" (Minutes March 10, 1956). The transfer occurred on October 1, 1956, in the middle of Nissen's directorship. Nissen was made a Professor of Psychobiology at Emory and the other postdoctoral staff members were named research associates of the new sponsor. A corporate board was established with President Goodrich C. White, Chairman; Henry L. Bowden, Vice Chairman; Boisfeuillet Jones, Secretary; Charles O. Emmerich; Morgan S. Cantey; L. P. McCord; and James C. Malone. Ewell E. Bessent, Controller of Emory, was appointed financial officer of the laboratories. Nissen commented that, "in effect, only the membership and officers of the Corporation were changed" (Annual Report 1957).

A new "University Committee on the Yerkes Laboratories of Primate Biology, Inc." was formed on the Emory faculty, including Howard M. Phillips, Chairman, J. A. Bain from Pharmacology, Geoffrey Bourne from Anatomy, W. D. Burbank from Biology, M. C. Langhorne from Psychology, A. J. Riopelle from Psychology, and J. W. Wiggins from Sociology. Bourne and Riopelle would later be Directors.

John F. Fulton, long of the Yale School of Medicine and a sometime critic of Robert Yerkes, was incensed when he heard the news. In a letter to Long he was especially critical of the administrators at Yale and Harvard: "You told me what a mess things were in at Orange Park and that Emory University, I take it with the approval of Whit G. of Yale and the President of Harvard, had sold Yerkes' life work and incidentally the dreams of those engaged in primate study throughout the world, for one dollar. When you told me this, I was so shocked that I could scarcely believe it" (Fulton 1958).

Funding

The funding situation during the Nissen-Peacock years fits within the broader trends for the history of the laboratories. In his annual report of 1956, Nissen wrote explicitly, "the major support of the Laboratories this year came from several federal agencies, instead of mainly from private philanthropic organizations as in the past" (1). As can be seen in Table 9.1, the total mean annual budget increased from $85,000 to $164,000, still half of what it would be under Riopelle. The trend toward decreased reliance on private foundation and university funding is clear (38 percent vs. 83 percent under Lashley), while reliance on all governmental sources increased (to 57 percent vs. 12 percent under Lashley). Lashley (1958e) wrote to Niko Tinbergen that "research money in the U.S. is almost begging to be taken." A summary of the sources of income during the Nissen/Peacock years is presented in Table 11-2.

Both the Rockefeller Foundation and the Carnegie Corporation had announced that their most recent grants, expiring in 1955, should be considered terminal (Annual Report June 1955). In fact, however, both continued through 1960, although Rockefeller support was decreased from $40,000 to $10,000 annually beginning in 1956. It is likely that a major factor in securing this support was the fact that Nissen was well respected as a scientist at both the Rockefeller Foundation and the Carnegie Corporation (Morrison 1955; Weaver 1955).

A major development was the appearance of a new private benefactor, the Ford Foundation. During his first year as director, Nissen received a five-year, $300,000 grant to support his program of research on the effects of early experience on the development of adaptive and maladaptive behavior in chimpanzees. The Ford project marked a shift from

Table 11-2
Funding for the Yerkes Laboratories of Primate Biology During Nissen/Peacock Fiscal Years, 1956–1959

Source	Amount	% of Total (rounded)
Private		
Rockefeller Foundation	$ 40,000	6
Carnegie Corporation	40,000	6
Ford Foundation	170,000	26
Wilkie Foundation	1,000	0
Fund for Neurobiology	2,000	0
Total	253,000	38
All Government		
Atomic Energy Commission	57,342	9
Public Health Service	66,664	10
National Institute of Mental Health	91,722	14
National Science Foundation	160,000	24
Total	375,778	57
Other		
Committee for Research on Problems of Sex	2,000	0
Sale of Animals	1,300	0
Rosalia Abreu Memorial	1,743	0
General Fund	4,472	1
Miscellaneous	20,772	3
Total	30,287	4
Grand Total	$659,015	100

studies of developmental effects on problem-solving behavior to the study of social interactions.

Clearly, the new director and the YLPB were in need of a single, core source of support, to which other sources could be added. They received this from the National Science Foundation in 1955. On October 8 and 9, 1954 the Board of Scientific Directors met with representatives of the NSF, including Marston Bates, Frank Beach, Harry Harlow, Louis Levin, W. J. H. Nauta, and John T. Wilson, in Orange Park. This paved the way for a three-year grant of $120,000, later increased to include a fourth year at $30,000. The grant was critical in serving as a bridge to keep the facility functioning with decreased Rockefeller support before base support was provided from the USPHS.

Other federal support was granted for specific projects. The USPHS granted what was supposed to be $30,000 for five years to support the developmental research and the U.S. AEC renewed its grant for 1955–56 for $19,570 (Annual Report June 1955).

Nissen secured a grant of $4,200 from the Committee for Research in Problems of Sex of the National Academy of Sciences-National Research Council for the period of one year ending June 30, 1955 (later extended) to study the interaction of innate and experiential factors in the development of sexual behavior in adolescent chimpanzees. He received a second grant, for $2,000, in 1955.

Other sources of funding included two small grants to Kinder for her continuing developmental work, monies from the Rosalia Abreu Memorial Fund, a small amount in the Yerkes Laboratories General Fund, and some small remnants of the Fels funding.

With the development of federal funding, there was a shift from general support for the Laboratories to targeted support for specific projects. This was directly counter to Lashley's approach. Well into retirement, he saw the handwriting on the wall and expressed his concern in a letter to Carmichael: "The major difficulty is probably the restriction of funds to special projects. I was able to give good men a chance to work independently on their own problems; now the staff is committed to projects in which they lack a personal interest" (Lashley 1957b).

Times were changing and the freedom that Lashley could afford his staff was becoming a thing of the past. The explosion of research that occurred with the influx of federal funds could not have occurred without them. One cost was exactly the kind of restriction that Lashley saw coming. It would get worse. Here, in a microcosm, we see yet another way in which changes in funding at the YLPB reflected changes in the nation at large.

Animals

The chimpanzee colony remained stable. When Nissen assumed the directorship he inherited a colony of sixty-six chimpanzees, fifty nine of whom were in Orange Park and seven were on loan. When Riopelle took up full-time residence in June 1959, there was a colony of seventy chimpanzees. For the first time in a while, new animals had been purchased: three jungle-born infants were acquired for use in the infant studies program so that their behavior might be compared with that of laboratory-born animals (Annual Report June 1959). Because of the difficulties in breeding the chimpanzees, Nissen (1957 Annual Report) noted that procedures were now in place to attempt artificial insemination. Otherwise, the health of the animals was generally good. New medications were found to deal with the intestinal parasites that had caused problems over the years.

Physical Plant and Equipment

The buildings required repeated repairs as the animals and climate were hard on them. With funds primarily from overhead allowances in the Ford Foundation grant, a new building, "the F-building," was constructed to house adults in twelve one-animal units. Each unit had an insulated compartment, one side of which could be fitted with panels adapted for use in assorted behavioral experiments (Annual Report June 1957).

The buildings occupied only a small portion of the land owned. Early in 1957 Clay County officials inquired about the possible purchase of forty acres of land for the construction of a public high school. The tract in question was the westernmost part of the property. Despite protests from Carmichael and members of the corporate board to state and local officials, the corporate board faced a choice between negotiation and condemnation proceedings. At its June 1958 meeting, the corporate board authorized negotiations for sale of the land (Annual Report 1958). Orange Park High School today is an active and sprawling institution. Few of its inhabitants are aware of the history of the property as part of a facility for chimpanzees.

Cooperative Efforts

The laboratories were involved in numerous cooperative activities with scientists at other institutions. Anatomical, fecal, and urine specimens were provided to some. Photographic material was provided for various publications. Classes from various schools, colleges, and universities around the state visited Orange Park. Many scientists from all over the world visited; visitors during this period included numerous alumni from the early years such as Glen Finch, K. L. Chow, James Gavan, W. N. Kellogg, and T. L. McCulloch. In a hint of things to come, during the 1957-58 year the Laboratories were visited by a committee, made up of Theodore C. Ruch, Harry Harlow, and Elsa Keiles, investigating possible sites for a national primate laboratory.

Life in Orange Park

Descriptions of life in Orange Park at this time suggest only gradual change from that during earlier periods. Janeth Carpenter (1993) quoted some of her earlier letters about her experiences. In June, 1956, she wrote: "The water in Orange Park was terrible, tasting of sulphur. We had to use bottled water for preparing coffee and soup, or the 'rotten egg' taste would overcome the taste of the food" (Carpenter 1993).

Even in the small community of Orange Park, the operant psychologists were generally isolated from the others. Marilyn Ferster Gilbert (1987) evaluated life at Yerkes as "strange but uneventful" (476). Gilbert (2002b) recalled that "the South—and Orange Park itself—seemed to us a wasteland." The Fersters raised son Billy in one of the Skinnerian "baby boxes" that provided excellent protection from the temperature changes and mosquitoes. This evoked great interest at the laboratory (Gilbert 2002a). She recalled that most of their neighbors were military families.

Lashley lived in Jacksonville. The area can be seen through Lashley's (1958e) eyes: "There are few intellectual contacts—a Little Theatre, a second rate symphony, a starting university graduating its first class next year, a good music school affiliated with the university, yachting on the river, and various social and golf clubs. Costs are no longer low here. . . . From July through Sept. temperatures may reach 95° in the afternoon but the nights are almost always cool."

Issues of race were beginning to become more visible. Gilbert (2002a) recalled that "we saw the separate water fountains when we went to Jacksonville, but little else." She remembered "another neighbor, originally from Savannah, came over one afternoon to tell me she had observed that the black garbage man had come to my front door, and I had invited him into my living room. I couldn't imagine why that mattered to her—until she explained that this was not done in the South" (Gilbert 2002a).

Some, like the Rogers and Menzel families, appear to have made many contacts and friends within the Orange Park community outside of the laboratories. Ginga Rogers even wrote a gossip column for the local weekly newspaper. Others, like Nissen, were more remote.

The strained social relationships that had characterized numerous phases of the history continued. In a diary entry, Warren Weaver (1955), of the Rockefeller Foundation, termed the Hayes-Nissen saga "Desire Under the Palms." He noted that his Rockefeller colleague, Robert Morrison, was left "with a vague feeling of uneasiness which has not wholly disappeared."

12
Research During the Years of Directors Nissen and Peacock

THE NISSEN-PEACOCK ERA WAS NOT AN UNUSUALLY PRODUCTIVE period for research, although some of it was notable in various respects. Nissen shut down the physiological and anatomical studies that had been prevalent under Lashley and returned to an approach more like that under Yerkes.

The three major areas of research were the program in operant psychology, the radiation program, and the infant studies program. It was the former that divided the research staff perhaps to the greatest extent in the history of the Laboratories. However, while the operant conditioners turned out a series of published reports, the radiation and infant studies programs brought fewer results to fruition at the time.

THE OPERANT PROGRAM

The Controversy

Operant psychology, or behavior analysis, had been developed by B. F. Skinner and his associates in the 1930s and 1940s and had been overshadowed by the learning theories of Clark Hull, Edward Tolman, and Edwin Guthrie for a number of years. In the 1950s, operant psychology was ascendant. In studies with animals, the subjects were generally food deprived and confined in a small, barren space that had been fitted with a device (the "manipulandum"), such as a key that could be pecked by a pigeon or a bar that could be pressed by a rat. Operation of the manipulandum produced a small reward, or "reinforcement." The manipulandum was attached to programming equipment so that the process could be automated. After a period of initial training, or "shaping," animals tended to respond in predictable ways. This afforded both precise control and precise measurement. Further, it was found that by arranging the programming equipment so that the reward was delivered after only some responses, interesting and predictable patterns of behavior could emerge. With "ratio" schedules of reinforcement, rewards were delivered only after a particular number of responses (fixed or variable). For interval schedules, only the first response after a particular interval of time (fixed or variable) had elapsed was rewarded. Other, more complicated, schedules of reinforcement were also developed (Ferster and Skinner 1957). The results were applied in a great variety of situations.

The events related to operant psychology that transpired in Orange Park need to be viewed in the context of what was happening across American psychology as a whole (Dewsbury 2003b). Behavior analysts were viewed from the outside as a closed group, isolated from most experimental psychologists primarily by the methodology that was adopted (Sidman 1960). For their part, many behavior analysts felt the need to distance themselves from the traditionalists because they believed that traditional methods were inadequate for the development of a true science of behavior and that operant research was not getting a fair hearing from journal editors and the program chairs for psychological meetings. Many outsiders perceived the behavior analysts as a group of "true believers."

Traditionalists favored group designs with tables of means, measures of variability, and inferential statistics. Behavior analysts preferred single-subject designs with control achieved through repeated measures. The operant single-subject designs were sometimes rejected for publication in the prestigious journals of the American Psychological Association. This rejection was an important factor in the founding of the *Journal of the Experimental Analysis of Behavior* in 1958 (Krantz 1972; Laties 1987), which increased the isolation between the groups. The operant period in Orange Park lasted just two years, from the summer of 1955 though the summer of 1957. It was a whirlwind event: the operant group swept in, dominated the research program for two years, and swept out again. This was a critical period of activity in the development of behavior analysis. Ferster and Skinner's *Schedules of Reinforcement,* published in 1957, was in preparation at the time. The meeting that resulted in the founding of *Journal of the Experimental Analysis of Behavior* was held a few months before Ferster's departure from Orange Park; all three Orange Park behavior analysts played a role in the founding of the new journal (Kelleher and Morse 1987). When the journal appeared the next year, Ferster was the founding editor. Thus, the Orange Park events transpired at a time when behavior analysts were very sensitive to discrimination against their approach and were in the process of organizing their own institutions in order to have appropriate outlets for their work. Falk (1994) remembered that "they felt that Henry [Nissen] really didn't want them to do what they were doing and there was a lot of tension." The personnel seemed divided into two camps. Charles Ferster was the leader of the operant camp. He came from Harvard and Skinner and was strongly committed to the operant approach. Roger Kelleher and John Falk were converted to the operant approach in Orange Park. Gilbert (2002a) recalled that "Charlie soon seduced John to behaviorism" and called her husband (Charles Ferster) "kind of a Pied Piper for behaviorism."

Nissen and others were attracted to the precision of the approach but were put off by other aspects. Surely the issues discussed above set the occasion for the tension between the two camps. The surface issues were more specific.

One issue concerned equipment. Peacock (1993) recalled that Ferster and Kelleher tended to look down upon the rest of the staff because they were using methods that appeared less sophisticated. For their part, the traditionalists were skeptical of the value of the expensive gear. Ferster frequently requested funds to purchase more equipment that had multiple uses. Peacock recalled that "Lashley said that he simply would not expend money for equipment that was not designed for a specific purpose. He said that in the first place if you can't do it with soda straws, twine, and rubber bands, it's probably not worth doing anyway" (Peacock 1993).

Those outside of the operant group were especially critical of the procedure of motivating animals by subjecting them to food deprivation until they reached 80 percent of their resting laboratory body weights. Recollections differ even today. Peacock (1993) recalled that Nissen "deplored the technique of starving down to 80% of body weight as a motivator for chimpanzees" and that "he and Ferster continually argued about that." According to Rogers (1994), "Ferster believed it was written in stone that if an animal's body weight was reduced to 80% they were going to work. And they would rather work than starve. But that wasn't true for chimps." According to Menzel (1994), "Nissen used to say that . . . that you try to starve them into submission and some animals would starve themselves to death. They'd just quit working." By contrast, according to Falk (1994), "the animals that were reduced in body weight, and concurrently on schedules, were bright eyed and seemed to have a purpose in life."

Nissen's concerns may have been justified to some degree. In his Annual Report for 1957, Nissen noted that "quantitative inanition, occurring in the course of experimental procedures, was probably at least a contributory cause in the death of one adult and one juvenile male." Rogers (1994) was less delicate and contended that Ferster had gotten scared by animal deaths and had thrown a lot of food in the cage of one of the chimpanzees. According to Rogers, the animal "ate everything in sight" and then upchucked, aspirated, and died.

When some animals failed to respond in ways characteristic of other species, accusations were made that staff members had been secretly supplementing the chimpanzee's diets. There are contradictory versions about possible sabotage of the operant conditioners' deprivation procedures. Some thought that Nissen would occasionally slip extra food into the animals' cages (Falk 1994). Gilbert (2002a) remembered that Ferster and Kelleher spent several nights watching the cages and saw nothing untoward. "Eventually, they decided that the chimps were eating acorns."

Skinner complained to Carmichael, who brought the charges to Nissen's attention (Carmichael 1957). Nissen defended his own conduct: "the statement is incorrect and misleading if it suggests that I have in any way interfered with, or attempted to regulate, the diets used with the subjects" (Nissen 1957b). He wrote that both he and Ferster agreed that the use of prolonged inanition was not consistent with the best aims of the program and that chimpanzees were not the ideal subjects for such research. Nissen concluded by affirming the value of the Skinnerian technique as applied elsewhere but not at the YLPB. Nissen wrote that "the statement . . . that the dietary restrictions necessary for Skinnerian techniques cannot be imposed on our chimpanzees is certainly misleading" (Nissen 1957b) without some elaboration. In some cases there had been more difficulty than with other species "in getting stable performances, uniform day after day" and "in holding weights constant from day to day" (Nissen 1957b). He suggested that the working spaces in which the animals had been tested were less insulated from distraction than in the traditional Skinner boxes. He added that newer boxes, built by Kelleher and Falk, had largely overcome this difficulty. Nissen agreed that, in some cases, the apes had been slow in making certain discriminations. "In one case there was as much responding without 'reinforcement' as with reinforcement" (Nissen 1957b). He attributed this, in part, to the use of commercial laboratory chow, a non-preferred food, as the primary reinforcement.

The factions disagreed regarding the treatment of the chimpanzees. Whereas it was common in Orange Park to call animals by name, the hard-nosed behavior analysts would not do so, using only a number

(Falk 1994). In addition, most at the laboratories did not like the way in which the operant group treated the animals. On the other hand, Ferster noted that punishment was used frequently in the day-to-day operation of the Laboratories where positive reinforcement might have been more effective.

As Ferster and Kelleher were leaving, Rogers wrote to Frank Beach: "I suppose that you have heard that Ferster is leaving. I can't say that I am sorry to see him go but it will do the lab no good. He has formed a small sphere of influence and sown enough dissension to make things pretty unpleasant. There is no research worth a damn but the stuff put out by the operant conditioning boys. With him depart the disciples." (Rogers 1957).

Skinner, in a personal memoir written upon Ferster's death, presented his view of Ferster's experience: "he went to the Yerkes Laboratories in Florida, where, unfortunately, he found the atmosphere uncongenial. (Tender-hearted colleagues frustrated his efforts to reduce chimpanzees to a satisfactory state of deprivation)" (Skinner 1981, 261).

The Operant Research

Despite all of the controversy, as Rogers suggested, the work of the operant conditioning group was probably the most productive done in Orange Park during this period. Several of the studies were published in the prestigious journal *Science*. The research included studies of schedules of reinforcement, timeout from reinforcement, conditioned reinforcement, and discrimination learning. Most of these studies appear to have been conducted successfully with chimpanzees maintained at approximately 80% of resting, laboratory body weight.

SCHEDULES OF REINFORCEMENT

With regard to the use of schedules of reinforcement, Nissen reported in his 1956 Annual Report that: "Intensive work with six adults and two young animals has demonstrated the feasibility of using this method with anthropoid apes. Various schedules of reinforcement have been explored, and in respect to most basic characteristics of responses the behavior of these primates has paralleled those of other animals. It is difficult at this time to interpret the differences found between chimpanzees and other species in certain details of behavior" (Nissen, Annual Report 1956).

Typically, the food-deprived chimpanzee was studied in a compartment with one or two telegraph keys mounted so that they could be operated easily. The rewards were generally alternated, with either a 50-calorie portion of whole-wheat cracker or Purina monkey chow provided on successive occasions. Typically, the animals were fed fruit, skim milk, and other supplements immediately after the conclusion of daily testing. Various complex schedules were used. These included chain schedules (Ferster 1958a) and concurrent schedules (Ferster 1957a, 1959).

TIME OUT

Having shown that chimpanzees respond appropriately in situations using positive reinforcement, Ferster (1957b, 1958b) studied control of behavior by punishment. Withdrawal of a situation in which positive reinforcement can occur constitutes a "time out." It is an aversive event. Chimpanzees were trained to respond for food reward on a variable interval schedule. They then were exposed to conditions in which the overhead light was turned off and the food magazine was disconnected from the response key simultaneously (a time out). Ferster noted that the time-out manipulation produced behavior quite similar to that which had been produced using shocks as an aversive event in pigeons. He likened the time outs to human situations with fines, disapproval, or incarceration.

CONDITIONED REINFORCEMENT

It will be recalled that during the Yerkes years at the Laboratories both Cowles (1937) and Wolfe (1936) studied conditioned reinforcement by training chimpanzees to work for token rewards that later could be exchanged for food. This is analogous, of course, to the human situation in which we work for money or other conditioned rewards. Kelleher conducted a similar research program using operant methodology. Not surprisingly, Kelleher (e. g., 1957) automated the process. In general, he found that the chimpanzees would work for tokens and the manner in which they did so depended on the schedule of reinforcement and whether they could exchange the tokens for food immediately or only after several had been accumulated.

CONCEPT FORMATION

Kelleher (1958) conducted an early study of concept formation. Two chimpanzees were faced with a stimulus board with an array of 9 small panels (3 x 3), any of which might be illuminated or dark. Different combinations of panels were illuminated on successive trials. Responses on a telegraph key were rewarded in the presence of some patterns but not others. In the first problem, all positive stimuli had in common the fact that the bottom row of panels was illuminated. For the second problem positive stimuli had in common that three panels, rather than two or four, were illuminated. Animals were trained with six positive and six negative arrays. Once the original patterns were learned, new arrays, not yet seen, were used to test generalization. If the animal had formed a "concept" of what the positive stimuli had in common, they should be able to generalize correctly from the familiar patterns to arrays not previously seen. They did this successfully for the first, but not the second, problem.

PERFORMANCE WITH ANOTHER REINFORCER

Falk (1958) conducted a study that did not require any of the food restriction or sophisticated equipment that proved so controversial. He noticed that, after testing in a delayed matching-to-sample study, a chimpanzee, identified as both Malcolm and No. 141 in his report, would throw "a tantrum" unless permitted to groom the experimenter's arm. Falk built a simple handheld plastic device that contained two patterns made of tape. Each day after the day's regular delayed-response testing, Falk conducted a series of trials in which Malcolm was to touch the correct pattern on the plexiglass device. If he did, Falk sounded a "cricket clicker" and presented his right arm for a 30-second period of grooming. The chimpanzee learned the discrimination, responding to the square when it was positive and to the cross-stimulus when conditions were reversed.

THE RADIATION PROGRAM

Part of the justification for the hiring of Ferster lay in the promise of the operant approach for the radiation program. The idea was that the operant techniques would provide a more sensitive test for effects of radiation on behavior than would the techniques previously in use. In the typical paradigm, an animal would be trained, then trucked to Oak Ridge, Tennessee for exposure to irradiation, then tested back in Orange Park for effects of the procedure on performance in the original task.

Radiation had no effect on the behavior of one of two animals trained on multiple schedules of reinforcement and exposed to 375 roentgens of total body irradiation from Cobalt-60 sources at Oak Ridge. The other, which died on the 28th day, showed no effect on the fixed-ratio schedule until death but a sharp rate drop on the 25th day (Nissen Annual Report, 1957).

Two other animals that had been trained on the concurrent fixed-ratio (FR) and variable-interval (VI) schedules were similarly irradiated. One animal showed only minimal changes after treatment; the other stopped responding on the VI lever completely immediately after radiation, whereas performance on the FR lever was unaffected. Nissen (Annual Report 1957) attributed the cessation of responding to motivational and emotional factors.

Research by Dowis and Kelleher, searching for delayed effects of irradiation using more traditional methods, also produced ambiguous results. Adults were tested a year after treatment; juveniles six months after treatment. In delayed response tests chimpanzees mastered spatial concepts readily but did poorly when new combinations of lights were used. Nissen (1957 Annual Report) stopped short of attributing these effects to radiation.

Attempts to breed irradiated females met with marginal success. In 1957, Nissen reported that one female that had been exposed to 400 roentgens in 1954 appeared to have become sterile for two years, after which she conceived. Parturition was difficult. A second female given the same treatment refused to mate with any of several males tested. Yet a third female, given 375 roentgens pre-pubertally became pregnant during the first cycle after her first menstruation (Annual Report 1957).

The following year, Peacock, writing the 1958 Annual report, noted that Nissen had suggested a pause in the five-year-old radiation program; that had been done. "The radiation studies program, therefore, has for the past year consisted of the maintenance of irradiated animals in optimal health, observations on their behavior in routine colony life, and the breeding of irradiated animals with each other and with normal animals insofar as possible" (Annual Report 1958, 3). No reliable effects of irradiation were found during these observations. Publication of these results would await the arrival of a new director.

The Infant Studies Program

Reports of the infant studies program during this period were long on methods and short on results. The basic design of the infant studies program was to rear chimpanzees under various conditions to examine the consequences for the development of behavior. Nissen noted that a new phase of the program was begun in 1956 when one subject, born on January 3, was reared in a "socially isolated, environment poor" condition (Annual Report 1956). His report was focused on methodology, with few new data to discuss.

By the next year, Nissen could report that at least two subjects were being reared in each of four conditions: monotonous environment, varied environment, social isolation, and social contact. For those experiencing a varied environment in social isolation, one was tested as "active" and the other as "passive." The active animal could operate switches that would produce various visual stimuli; the passive animal had identical, but inoperative, switches so that it experienced the pattern of visual stimuli resulting from responding by the active animal (i.e., yoked control).

As was often the case, Nissen and Lashley did not see eye to eye; Lashley thought little of the program. Lashley (1958e) wrote to Niko Tinbergen, a prospective successor to Nissen, that "at present the entire staff is engaged on a single project: rearing of chimp. Infants under diverse social conditions with tests of effects on subsequent behavior. I suspect that you would want to revise it fundamentally or junk it-I would certainly do so."

The brief Nissen-Peacock period was a time of challenges. As these were being met, the research program, with the exception of the operant contingent, produced few tangible research reports during this relatively brief period in the YLPB's history.

13
The Riopelle Years: Part I

THE NEXT PHASE IN THE HISTORY OF THE YLPB WAS ONE OF SWEEPING changes. The full flowering of federal funding for research brought an active, productive, and savvy group of scientists to Orange Park. Although he was not always perceived that way, director Arthur J. Riopelle was the ideal man to take the laboratories into this new era. However, that same development of federal programs would cause much strife between personnel at Orange Park and in Atlanta, with the final result that the end of the Orange Park facility was in sight.

In his first year, Riopelle would continue programs already in force. However, he would gradually add personnel, reshape the character of the YLPB, and make it more competitive in the new funding environment. By June 1961 he had a National Heart Institute (NHI) grant in place, the personnel count up to 50, and the chimpanzee count up to 83. Riopelle again diversified the range of species housed at the YLPB. The process of redevelopment ended when Emory stepped in to take full advantage of the new availability of federal funding. The few years under Riopelle saw rapid development but the dominant theme was the struggle over the future of the laboratories as an entity in Orange Park or after removal to the Emory campus.

FINDING A NEW DIRECTOR

The Search

There had been concerns about Nissen's directorship from the beginning. Ten days before Nissen's death, Charles T. Lester, of the Emory Graduate School, wrote to Carmichael that "it is the feeling of our faculty and many other people not connected with Emory University . . . that the program at Yerkes laboratories needs to be expanded and that it needs the direction of a man with broader interests and more vigor than Dr.

Photograph of Arthur J. Riopelle, fourth director. Photography by the author.

Nissen" (Lester 1958a). Lester suggested a list of six names of possible candidates for the directorship. The six were ranked in order: Harry Harlow, Frank Beach, Donald Hebb, Austin Riesen (then of the University of Chicago), Donald Meyer, and Arthur Riopelle.

After Nissen's death, the young Lee Peacock was regarded as strictly an interim director, not experienced enough for the job as permanent director. The job of replacing Nissen was undertaken in earnest.

Many other names were placed in the hopper for consideration as the new permanent director. An unsigned memorandum in the Carmichael Papers of the American Philosophical Society, possibly from Carmichael, summarized the results of a conference with Frank Beach, Harry Harlow, and Heinrich Klüver at the meeting of the Society of Experimental Psychologists and listed eight suggestions. Lashley (1958b) addressed his suggestions to Dean Arthur Richardson in the medical school at Emory in a letter written about a week after Nissen's death. He lobbied for an Emory faculty appointment with tenure to help attract the kind of director Lashley envisaged. He listed numerous candidates, some of whom were clearly unavailable. Lashley recommended YLPB alumnus James H. Elder because "he has done good work on sex physiology and behavior at the Laboratories." For one interested in "broader studies of instinct," he wrote that Niko Tinbergen, the Oxford ethologist, "should be the first choice for the field." Of Harlow he wrote "should be good administrator and money raiser, but research is second rate." Lashley (1957a) had been critical of Harlow earlier as well, both for what he regarded as second-rate research and because "that little bastard has tried to stick his nose into the affairs of the labs."

The kind of individual to be chosen would affect the kind of research to be done. Long favored cancer research because of the funding available. Bourne suggested taxonomic, genetic, and nutritional studies. Several other possible directions were considered. In the end, the Board discussed the qualifications of thirty-one possible candidates and voted to ask six individuals in rank order: Hebb, Niko Tinbergen, W. D. Neff, Jose Delgado, Elliot Stellar, and William C. Osman-Hill (Board Minutes May 10–11, 1958).

The process continued with Lester writing letters of invitation. He wrote to at least Hebb (Lester 1958b), Tinbergen (Lester 1958d), and Neff (Lester 1958c) and perhaps others. In the end, the new director did not come from that list.

Hebb wrote to Carmichael of his two primary reasons for declining; the first was prescient. Hebb noted that Lester had written that control of the YLPB was in Emory's hands with advice from the Board of Scientific

Advisors. Hebb added, "I got the feeling that he was approaching the matter from the point of view of forwarding Emory's welfare and interests. These are not necessarily identical with the Laboratories'; and so a Director might be faced with the further difficulty of coordinating the two, and having to get involved in Emory politics as well as running the Laboratories" (Hebb 1958). Hebb was also concerned that there were just two comparative psychologists on the board whereas he saw the future of the YLPB as lying in comparative psychology and the new director as benefitting from the advice of comparative psychologists.

Tinbergen (1958) was interested enough to request further details, which Lester (1958e) provided. However, he declined as well. Neff was sufficiently interested so that he visited both Atlanta and Orange Park. He wrote, repeating some oft repeated themes, that "to me, the most discouraging aspect of accepting a job at Orange Park is having to live there" (Neff 1958a). In a second letter, Neff (1958b) suggested to Lester that facilities should be added for animals other than chimpanzees, the behavioral emphasis should be continued but with work in other fields added, more accommodations should be made for visiting investigators, additional equipment should be purchased, the salary for the director should be increased from $12,000–$15,000 to approximately $20,000, more competent staff members should be attracted, and the facilities should be expanded. In declining, Neff appended an annotated list of his own suggestions for the position of Director.

With the top choices declining, it appears that Richardson was the one who began to push for Riopelle. In an August 16 telegram to Carmichael, Corner (1958) suggested favorable consideration, although he did not seem to know Riopelle well. By October 24, Carmichael (1958) wrote to Richardson that he had been in contact with all available members of the board and they agreed in recommending Riopelle. By November 12 Riopelle had been offered the job and wrote to Emory President S. Walter Martin concerning several issues that should be addressed before he made a final decision. By February, Lester (1959) informed the Rockefeller Foundation officials that Riopelle had accepted the position. Arrangements were made so that Riopelle could complete his obligations at the Psychology Division of the U.S. Army Medical Laboratory while serving part-time as the new director. Riopelle took up residence in Orange Park on June 1, 1959, twenty-nine days after Lashley's death and twenty-nine days before his first annual report was due.

The Fourth Director: Arthur J. Riopelle

Arthur J. Riopelle was born April 22, 1920 in Thorp, Wisconsin, a small town of about 400 people. He was one of six children whose father was a physician and surgeon. He received his bachelor's, master's, and PhD degrees from the University of Wisconsin in 1941, 1948, and 1950 respectively. During the war, Riopelle served in California and Europe with duties focused on assessing the adequacy of various tests as predictors of performance for bombardiers, navigators, and pilots (Dukelow 1995). In 1942 he married Mary Jane Astell, a fellow student. The Riopelles would have three children, all of whom would work in medicine or psychology (Dukelow 1995).

Riopelle's PhD research, which dealt with sensory processes, was supervised by David Grant along with Fred Mote. In addition, he did some testing for Harlow W. Ades of Emory University. It was Ades who liked his work and arranged an appointment on the Emory faculty. Riopelle served as an assistant and associate professor of Psychology on the Emory University faculty from 1950–57. It was also through Ades that Riopelle became interested in primates and the effects of radiation. Riopelle was able to build a small primate laboratory at Emory. He then moved to become the director of the Psychology Division of the U.S. Army Medical Research Laboratory in Fort Knox, Kentucky during 1957–59. After serving as the director of the YLPB during 1959–62, Riopelle became the founding director of the Delta Regional Primate Research Laboratory of Tulane University in Covington, Louisiana for 1962–71. He then served on the faculty of Louisiana State University and retired in 1989. Riopelle authored over 150 articles in professional journals.

Administration

The Board of Scientific Directors

Riopelle inherited a board of scientific directors consisting of Leonard Carmichael (chairman), Geoffrey H. Bourne, George W. Corner, Frederick L. Hisaw, C. N. H. Long, Arthur P. Richardson, and William H. Taliaferro (Vice Chairman). When Carmichael announced his wish to step down, Hisaw was elected chairman and Richardson as vice chairman at the March 1961 meeting.

The Board continued to meet in Orange Park each spring with additional meetings as required. Although this was a non-threatening group, not out to conduct any really critical evaluation, extensive preparations continued to be made at the YLPB each year. The laboratory was cleaned, floors were oiled and polished, and there was great attention to the condition of the facility. The Orange Park meetings typically began with a tour of the grounds and staff presentations, the "dog and pony shows." The Board then conducted its business related to research, funding, physical plant, animals, and the like.

Draper (1994) remembered these visits as being great fun for the staff; the research presentations occurred in a non-threatening environment and the board was quite supportive. He recalled several occasions on which the chimpanzees soiled the board members' new silk suits with their usual antics. On another occasion, the meeting was held outdoors on the laboratory grounds. It was rapidly moved indoors after a coconut fell from a tree with a thud, barely missing one of the board members.

The Board and Emory

In the early years, Yale and Harvard had allowed great freedom in the administration of the laboratories. Emory, by contrast took firmer control. At its June 1960 meeting, Carmichael read a letter from Emory President Martin informing Carmichael that a new advisory committee had been appointed to advise him on matters related to the YLPB and consisting of the two vice presidents of the university, the deans of the graduate school and medical school, the director of the division of basic health sciences, and the business manager of the university. President Martin added that the committee was not to conflict with the activities of the board. At the same meeting, however, Dean Richardson indicated that Emory would not approve a purchase of additional land to provide a buffer from the growing community.

Various other Emory officials attended board meetings. At a special meeting held in December 1962, Emory official Orie Myers requested that all expenditures from the main USPHS grant for travel, personnel, equipment, construction, or renovation in excess of $25 receive prior approval from his office in Atlanta.

Personnel

Scientific Staff

As Riopelle took up residence he found just four continuing staff members: Davenport, Menzel, Peacock, and Rogers. Peacock soon resigned;

Riopelle had lowered his salary because he believed that Peacock was now free of administrative responsibilities and maintained a minimal research program. When the laboratories were moved, Davenport and Rogers moved with them; Menzel joined Riopelle in Louisiana.

Riopelle sized up the financial situation and saw that the RF was about to terminate funding and that Emory would not contribute any funds to the laboratory. He reasoned that he needed to add new members of the scientific staff who could generate funds from the increasingly significant Federal funding grants programs. Like previous directors, he set about rebuilding the staff (see Table 13-1).

Riopelle's first three appointments were just the kind of people needed in the new environment. William Mason, perhaps the key addition, was a Stanford PhD and a colleague of Harry Harlow's at Wisconsin. In March 1962, when Riopelle felt that he needed additional administrative help, Mason was named the assistant director. The soft-spoken and reflective Mason remained a leader in primate research from his days in Wisconsin into the twenty-first century.

John Michael Warren completed his PhD with Harlow at the University of Wisconsin in 1953. He joined the YLPB research staff in August 1959 after serving at the University of Oregon and Stanford University but remained in Orange Park for just a year. Warren worked primarily with cats while at the YLPB; he did little with the chimpanzees. Although Warren was the kind of scientist needed in the new environment, because of his short stay, his success came elsewhere.

Irwin Bernstein, a University of Chicago PhD who had worked with Riesen, joined the staff as a research associate on New Year's Day, 1960. He moved with the YLPB to Atlanta, remaining full-time on the staff until 1967 when he joined the faculty of the University of Georgia. He would function both as a fine researcher and as something of a conscience of primatology.

Several other key staff members complemented these three. Gershon Berkson, a psychologist with a primary interest in developmental disorders, became a USPHS postdoctoral fellow with Mason in 1960. Although he originally went to Orange Park to study the development of the EEG in chimpanzees, he soon changed his focus to work in the infant studies program. He had a special interest in the stereotyped rocking that occurred in some animals—an interest he still holds. Ronald Schusterman also joined the staff in 1960. He first became interested in work in Orange Park when Winthrop Kellogg, a Yerkes alumnus then at Florida State, brought his students for a tour of the facility. Schusterman worked with choice behavior, learning sets, and food sharing and also did some work with gibbons and with retarded children at the Sunland Center in

Table 13-1
Staff Members During the Riopelle Years with Birthplace, Birth Date, PhD Information, and Approximate Years of Service

Gershon Berkson. b. June 3, 1931, Jerusalem, Palestine. PhD Vanderbilt University, 1959. "A Study of Reaction Time and Duration Threshold in Familial Mentally Deficient and Normal Adolescent Boys." Orange Park 1960–63.

Irwin S. Bernstein. b. July 11, 1933, Brooklyn, NY. PhD University of Chicago, 1960. "The Utilization of Some Visual Cues by Primates in an Oddity Problem Situation." Orange Park 1960–65.

Orlando M. Daumy. b. 1902, Cuba. MD University of Havana Medical School, 1940. Orange Park 1959–63.

William A. Draper. b. September 27, 1928, Toronto, Ontario, Canada. PhD University of Oregon, 1961. "A Behavioral Study of the Home-Cage Activity of the White Rat under Normal, Deprivation, and Drug Conditions." Orange Park 1961–66.

Frances L. Fitz-Gerald. b. March 28, 1938, Wichita, Kansas. PhD McGill University. 1964. "The Effects of Drugs Upon Stereotyped Behavior in Young Chimpanzees." Orange Park 1961–65.

Naosuke Itoigawa. b. January 15, 1935, Hokkaido, Japan. PhD Osaka University, 1964. Orange Park 1961–63.

William A. Mason. b. March 28, 1926, Mountain View, California. PhD Stanford University, 1954. "The Effects of Hypophysectomy and Electroconvulsive Shock on Thermal Preference and Nesting in Rats." Orange Park 1959–63.

Lawrence R. Pinneo. b. February 11, 1928, Denver, Colorado. PhD 1962, McGill University. "Effects of Induced Muscle Tension During Tracking on Level of Activation and on Performance." Orange Park 1962–63.

Sue V. Saxon. b. February 13, 1932, Brooksville, Florida. PhD Florida State University, 1963. "Test Profile Characteristics of Selected Behavior Pattern Groups of Freshman Women Residents on the Minnesota Counseling Inventory." Orange Park 1960–61.

Ronald J. Schusterman. b. September 3, 1932. PhD Florida State University, 1961. "The Use of Strategies in the Decision-Behavior of Children, Chimpanzees and Monkeys." Orange Park 1960–63.

Narsingh Sharma. b. July 13, 1930, Jalaun, India. PhD University of Georgia, 1962. "Response of the fowl (*Gallus domesticus*) to seven species of Parenterally Administered Coccidal Organisms." Orange Park 1962–63.

Johannes Tigges. b. July 7, 1931, Rietberg, Germany. PhD University of Münster, 1961. Orange Park 1962–63.

John Michael Warren. b. September 6, 1925, Chicago, IL. PhD University of Wisconsin, 1953. d. July 4, 1991. State College, PA. Orange Park 1959–60.

Gainesville. He later became well known for his work in cognition and language learning in sea lions. Naosuke Itoigawa from the Japan Monkey Center and Kyushu University was one of several international scientists at the facility. He had been working with Japanese monkeys since 1957. In Orange Park he participated in a number of projects, including a study of reactions to mirror-image patterns, before returning to Japan. Orlando M. Daumy, MD, a 1959 refugee from Cuba and *persona non grata* under Fidel Castro, joined the staff as a Research Associate in February 1961. In Orange Park, he worked on the radiation program and collected normative physiological data on the animals. William A. Draper, Frances Fitz-Gerald, Lawrence R. Pinneo, and Johannes Tigges were among the other important contributors to the research effort.

Perspective on the Scientific Staff

Rogers and Davenport worked somewhat in the tradition of Henry Nissen. They were diligent in conducting fine research and really knew and cared about their animals. However, they were less concerned than others with publishing data and getting grants and appeared to lack the motivation to truly excel in the new environment. Riopelle's new staff members, Mason, Warren, Bernstein, for example, had a very different approach. Times were changing and they recognized that while diligence was an important first step, one needed to be visible through publications. The future of the institution depended on increased funding through individual grants and publications. One had to be, at least to some degree, also a publicist and entrepreneur. One might view it as a shift from a Nissen mentality to more of a Harlow mentality. This shift was occurring during Riopelle's brief tenure as director. It is fascinating to speculate how this might have gone if Riopelle and the laboratories had stayed in Orange Park.

Technical and Service Staff

As the scientific staff was growing, so was the support staff. Peacock's 1958 annual report listed 19 members of the technical and service staff. Riopelle's 1961 Annual Report listed 57. Stalwarts included Office Manager Murray Townsend, Superintendent J. G. Townsend, Administrative Assistant Mildred J. King, Technical Assistant Arthur Chester, and Caretaker Annie Jones. Several spouses of YLPB scientific staff members served part-time on the support staff. As in previous years, most of the lower-level positions, caretakers, laborers, and the like, were filled by African Americans; it was difficult for them to advance through the ranks. These staff members were essential to the success of the laboratories. A "people's history" approach would tell their stories, not just the story of the visible leaders of the laboratory. Unfortunately, their stories are not recorded as are the careers of scientists and so this essential part of the laboratory culture is neglected.

Funding

For the purposes of analyzing funding, I treat the Riopelle years as running from July 1, 1959–June 30, 1962. A primary source of data was annual reports, which were presented each year through June, 1962. The last report contains only financial data and was prepared by Murray Townsend. I have been unable to locate a report for 1963. The final two reports of the Orange Park era provide no financial data.

As can be seen in Table 9-1, the total budget for the reporting years 1960–62 were $1,163,234, or $387,745 per year. That was over twice as large as during the Nissen/Peacock years and nearly six times as much as during the Yerkes years. That the budgets were growing will come as no surprise to the reader.

The effects of inflation must always be remembered in such comparisons. According to the inflation calculator of the U.S. Bureau of Labor Statistics, the purchasing power of $100 in 1930 required $180.84 in 1962. Thus, although the average budget under Riopelle was roughly six times that under Yerkes, it was probably more like tripled in real dollars. More precisely and taking the most extreme years, Yerkes' 1931 budget of $40,000 increased to Riopelle's $395,613 by a factor of 9.9 but an inflation-adjusted factor of approximately 5.5.

The distribution of income sources shows how dramatically patronage for science both in Orange Park and in the country, had changed since Robert Yerkes' term. Private foundations had provided 79 percent of Yerkes support; it was just 13 percent under Riopelle. University support, which provided 16 percent of Yerkes' budget, was non-existent under Riopelle. Emory provided no funding. Fully 86 percent of Riopelle's income was from federal sources; Yerkes had no such support. These data are broken down further in Table 13-2.

As noted earlier, with increasing budgets came a shift in expectations. Much of the support was for specific projects rather than general funds.

The last funding from both the RF and the Carnegie Corporation was received in 1960; each contributed $10,000. The remainder of private support during this period was the continuation of Ford Foundation funding for the Infant Studies Program.

Table 13-2
Funding for the Yerkes Laboratories of Primate Biology During Riopelle Fiscal Years, 1956–1959

Source	Amount	% of Total
Private		
Rockefeller Foundation	$ 10,000	1
Carnegie Corporation	10,000	1
Ford Foundation	130,000	11
Total	150,000	13
All Government		
Atomic Energy Commission	45,542	4
Public Health Service	1,500	0
National Institute of Mental Health	126,470	11
National Institutes of Health	182,532	15
National Science Foundation	33,500	3
National Heart Institute	472,500	41
U. S. Army Surgeon General	135,687	12
Total	997,731	86
Other		
Rosalia Abreu Memorial	1,459	0
General Fund	13,044	1
Miscellaneous	1,000	0
Total	15,503	1
Grand Total	$1,163,234	100

The National Science Foundation, which had provided $40,00 per year during 1956–59, provided no funds in 1960, minimal funding in 1961, and $33,000 for 1962. The Atomic Energy Commission continued support of the radiation project for the first two years under Riopelle.

Some of the funding from the USPHS, NIMH, and NIH is difficult to sort out because of the hierarchical organization of the institutes. Different reports might list the same funding under different headings. I have used the sources as listed in the annual reports. The fundamental support was a grant of $150,000 plus overhead per year for a period of ten years from the National Heart Institute. That provided general support for the laboratory infrastructure and a greatly increased sense of financial security. During 1959–60, three additional grants from the NIH were received. (The heart grant H-56791 is listed under NIH in the financial summary statement.) Riopelle also received a contract from the U.S. Army Office of the Surgeon General.

Both the 1961 and 1962 annual reports list fifteen different pots of federal money from which the YLPB was receiving support. The bulk of these accounts were from the NIH. Three of the sources were Fellowship Supply Grants of less than $1,000. For 1962, seven different NIH grants were listed. It is not clear for some of these which was for support of which projects.

Bookkeeping was more complicated than in previous years. Great care had to be taken in preparing the budgets so that the scientists paid through particular grants were actually working on the appropriate projects (Board minutes June, 1960).

Whereas simpler financial summaries sufficed in the earlier years, breakdowns of expenditures for the various overlapping grants were necessary in the annual reports of this period. The expenditures during the Riopelle years, as recorded in the three annual reports, are given in Table 13-3. As can be seen in the table, nearly two-thirds of the expenditures were for personnel costs. These costs rose by 45 percent in just two years. During that same time, expenditures for scientific supplies and apparatus rose by a factor of 5.6. Travel funds more than tripled. The costs of animal maintenance (animals, food, and supplies) appear re-

Table 13-3
Expenditures for the Yerkes Laboratories During the Years 1960–1962 under Director Riopelle

Category	1960	1961	1962	Total
Salaries	$122,440	$157,959	$177,890	$458,289
Wages	19,588	32,076	36,307	87,971
Social Security	3,473	2,498	2,807	8,778
Annuities	1,658	4,891	5,768	12,317
Workmen's compensation	1,905	4,452	2,862	9,219
Total personnel	**149,064**	**201,876**	**225,634**	**576,574**
Scientific supplies and apparatus	8,744	22,052	48,994	79,790
Animal maintenance	14,050	26,842	33,859	74,751
Plant insurance	1,496	1,272	1,732	4,500
Plant maintenance	16,966	54,888	39,310	111,164
Library & publications	1,095	3,658	1,645	6,398
Travel	3,436	6,537	11,125	21,098
Special services & Misc.	527	527	518	1,572
Grand total	**$195,378**	**$317,652**	**$363,817**	**$875,847**

markably small, accounting for just 8.5 percent of the total cost of running the station. They rose by a factor of 2.4 over two years.

Physical Plant

The alterations of the physical plant during the Riopelle administration were relatively minor. Nevertheless, the plant was beginning to show some effects of age and use and a host of repairs were in order. With space at a premium, much of the caretaker's house was converted to laboratory space. The old badminton and tennis courts, which were outside of the main fence but inside a secondary fence, were used as bases for constructing large monkey compounds where animals in groups could be observed. Perhaps these changed functions were symbolic of the changes that were occurring. The older, more casual way of life at the laboratories, with badminton and tennis, was giving way to a newer, more competitive, and intensive approach. Riopelle finally introduced steam cleaning to the facility.

Animals

Riopelle's brief term as director saw increases in both the number of animals at Orange Park and their diversity. Riopelle inherited a colony of seventy chimpanzees. In April 1960, the YLPB received a shipment of twelve rhesus and four dourocouli monkeys; in June they received eight gibbons. In addition, Warren maintained sixty-two cats and four monkeys for his research.

As Riopelle extended the interactive relationships with other institutions, the numbers became somewhat more complex as there were increased loans of animals to and from the laboratories. As of March 20, 1962, the colony in Orange Park included ninety-seven chimpanzees, and gorilla that had been purchased, twenty-five monkeys, and ten gibbons.

Miscellany

The assorted other activities of the YLPB staff continued and were expanded. Riopelle believed strongly in opening up the laboratories to the extent that was compatible with ongoing research and in fostering cooperative ventures with scientists from a range of institutions. Staff members attended and made presentations at the usual assortment of professional meetings. In addition, they gave talks at various universities and civic organizations and served on various committees and panels.

Cooperative ventures were developed with the University of Florida, the University of South Carolina, the Lockheed Aircraft Corporation, Emory, Wayne State University, Columbia University, Yale University, the National Institutes of Mental Health, the U.S. Naval School of Aviation Medicine, and the Sunland Training Center in Gainesville. Samples and data were provided to colleagues around the world. In his 1961 report, Riopelle noted that the laboratories had on hand over 1,000 samples of blood fractions available for distribution. There were many visitors; during the 1959–60 reporting year alone, over 217 people visited the laboratories.

Life in Orange Park

It had now been thirty years since the YLPB had been founded. Much had changed in society and in the laboratories themselves. Orange Park was growing and beginning to infringe on the YLPB property. Riopelle (1959c) wrote to Ada Yerkes that "Orange Park is growing like kudzu." He regarded the sale of property for the high school as foretelling hundreds of homes in the area. This turned out to be an underestimate and was one factor of concern if the facility should remain in Orange Park.

In spite of the growth, much about life in Orange Park had not changed to anything like the degree that it has since the laboratories were relocated. The portrayal of life in Orange Park presented by my interviewees from this era was remarkably similar to those from the earlier periods. Overall, most interviewees liked living in Orange Park, which they perceived as quiet and somewhat rural, and remembered their experiences fondly. There were some negatives, however.

Complaints

The same basic complaints as in earlier years were expressed. Florida is hot in the summer and there was little air conditioning. Schusterman (1993) recalled sometimes going to a movie or to the Greyhound bus station to sit in air conditioning. Berkson (1996) recalled concern about snakes as he had a two-year-old child. Johannes Tigges (Tigges and Tigges 1993) was frustrated by the cockroaches and his inability to catch them as they disappeared behind the molding of his house.

Hepatitis

A major problem arose in March 1960 when six African-born chimpanzees arrived in the laboratories. Seven members of the staff, all of whom were working closely with the new animals, and a girl visiting from Tallahassee, contracted infectious hepatitis within a month. There was concern that Davenport might die. After a thorough review by the Florida State Board of Health, all possible sources except the animals were eliminated. There appeared to be subclinical hepatitis in three of the animals. Fortunately, all of the affected employees were back at work by June 30.

At the Laboratory

The laboratories themselves provided a pleasant working atmosphere in which senior staff members were supportive of their younger colleagues. Berkson (1996) recalled the atmosphere of the library. Although small, this had been the library that Yerkes and Lashley had used and he was moved by their spirit there. Like others, Draper (1994) recalled a diligent work ethic but with a lack of pressure. There was frequent and open discussion of research projects.

Anecdotes from this time, as from previous epochs, often concerned escaped animals, the throwing of feces, and the spitting of water by the chimpanzees. One story was told by several individuals and obviously made a great impression on them. Occasionally a chimpanzee would escape when a cage had not been locked properly. The standard procedure was for individuals to blow whistles, signaling the escape, and then, for safety, go into a building until the escaped animal quieted down. On one occasion chimpanzee Julius escaped and the standard procedure was followed. Riopelle, however, entered the grounds with a pistol with which to scare Julius. Julius picked up a piece of bamboo and started to charge Riopelle. The gun did not fire and Julius went right past Riopelle; Riopelle then retreated to a building. The situation was resolved when one of the more experienced caretakers offered Julius a drink of Coca-Cola laced with a sedative. Julius drank it, went unconscious, and was carried back to his cage.

Another story that provides richness regarding life in the South at that time concerns transportation of chimpanzees to Tennessee for irradiation W. A. Draper made an audio tape for my use; I quote verbatim from our transcription of it:

> [a] brown truck was used to go to Oak Ridge, Tennessee and invariably these two fellows would take them up in the late evening and night and arrive there the next morning to do whatever testing they were going to do at Oak Ridge. And of course they had to drive up through Georgia and up through South Tennessee and eventually up into Oak Ridge. And it was moonshine country of course . . . One night the two fellows were driving along and the Tennessee State Police stopped them. Figured they were moonshiners. And they made the fellows get out, and they said, searched them I guess. They said open the back of that truck, we want to see what's in there . . . The fellows said you open the back of the truck yourself. So he opened the back of the truck and these two chimps which were adult chimps were in there and let out one unholy scream because they were nervous themselves and they were black, it was pitch black and dark. And so the trooper slammed the door and he said "Buddy, I don't know what you got in there, but I want you to get them out of my county and don't ever come back here again."

Staff Socializing

Outside of work the staff members socialized among themselves a good bit. Most staff workers lived in or around Orange Park. There were many parties, apparently averaging about one a month. Berkson (1996) and others described crabbing and shrimping. Itoigawa (1994) recalled making sukiyaki and tempura, as well as eating raw shrimp. There were also periodic oyster feasts and other parties at the beach.

Several alumni of the YLPB recalled stories of Penrod, a rhesus monkey hand-reared in the home of Irwin Bernstein and his wife. Penrod was not especially friendly to most humans and would use open-mouth threats directed at visitors who looked at him. Bernstein could control the animal because he was perceived as dominant. On one occasion, Penrod attacked Sheila Bernstein and the home-rearing ended (Schusterman 1993).

Race

The South at this time in its history was in the last throes of overt segregation. Both Helen Warren (1994) and Bernstein (1993) recalled segregated drinking fountains in Orange Park but not at the laboratory. Bernstein noted that because of the old plantation roots of Orange Park, black families were more scattered throughout the town than in many other communities. Draper (1994) provided the most detail in this regard. Like most northerners at the laboratory, he was a scientist with no wish to create a revolution but was sensitive to injustice. His approach was to make the best of the situation and make small inroads where it was feasible. At the YLPB there was no overt enforcement of segregation; all knew how things were.

Draper recalled an instance when he was sitting at the lunch counter in the local drugstore talking with a white man. A friend, a black animal handler, came up and stood behind them socializing. Draper invited the handler to have a cup of coffee. The handler took the cup from the counter and remained standing behind them while drinking the coffee. Draper, new to the South, committed the faux pas of asking why the handler did not sit down; he got no answer. Draper then stood with him and only went back to his seat when the handler left.

On one occasion, a black couple was invited to a laboratory party—a highly unusual event. Nothing was said at the party but there were comments the next morning at the laboratory. There was no animosity; it was just made clear that this was not how things were done in the South (Draper 1994). Some uneasiness was created when a black summer student was hired as an assistant rather than as an animal handler.

Berkson (1996) told another story regarding the naïveté of the northern staff members regarding southern segregation. Soon after moving to Orange Park, his wife brought a load of laundry to the local laundromat. She interpreted the "whites only" sign as an instruction related to the separation of white from colored laundry, not people.

The beaches in St. Augustine were segregated. On one occasion, Draper took his family to observe a Ku Klux Klan pro-segregation demonstration there.

Community Perceptions of the Laboratories

Residents of Orange Park continued to be curious about the laboratories. Tales about efforts to cross-breed a human and an ape continued to circulate. One of the rumors was that that was why so many blacks were hired as technicians. There were still stories of Boy Scouts camping out around Doctors Lake, hearing the calls of the chimpanzees, and telling scary stories late at night.

Riopelle tried to change these perceptions by opening up the laboratory to a greater extent than had his predecessors (Riopelle 1994). He invited the mayor and his family out to tour the laboratories and explained the research to them. Groups of children were invited in; the theory was that they would quell some of the fears of their parents. In May and June 1961, the local CBS-TV affiliate aired a program on the thirty-one-year history of the YLPB in its "Project Four" series. This would have been unthinkable under Yerkes.

Overall Experience

Overall, Draper (1994) recalled his years in Orange Park as "some of the happiest and yet most poignant years of my life." Itoigawa (1994) called his days in Orange Park "most exciting, fruitful, and pleasant" and called the YLPB "THE most important institution in the history of psychobiology." Schusterman (1993) believed that "everything, any kind of science I've done since then has been influenced by my experience at the Yerkes laboratories."

14
The Riopelle Years: Part II

Research

At his initial meeting with the Board of Scientific Directors as the new Director, Riopelle outlined a four-pronged program for the concentration of research for the YLPB program:

> 1. Analysis of primate behavior, including problem solving, perception, social behavior, and free-ranging behavior both from the standpoint of theoretical psychology, and from the standpoint of contribution to the comparative studies of behavior.
> 2. Study of physical anthropology of primates, including comparative anatomy, physiology, embryology, endocrinology and hematology.
> 3. Study of brain functions through neurophysiological and behavioral means.
> 4. Study of the behavioral effects of insults and unusual environments, such as: radiation, drugs, cortical ablations, sensory deprivations, immobilization and intense stimulation (Board minutes May, 1959, 7).

The realized program deviated somewhat from this plan with studies of behavior, especially of the development of behavior, being the most productive. Some of Riopelle's plans, particularly as related to items 3 and 4, never really developed fully. It should be noted that, because of greater continuity of staff during this period, some of this research transcended particular directorships.

Studies of Basic Behavioral Patterns

Bernstein (1962) compared nest building in captive-born and wild-born chimpanzees. Whereas all of the wild-born animals built nests, fewer than half of the captive-born animals did so. Although they all used similar motor patterns, the nests built by captive-born animals were generally more crude than those of wild-born chimpanzees. The results suggested an important learned component in the development of nest building.

Bernstein (1964) revived the Yerkes tradition of field work, following up Carpenter's earlier work with an observational field study of howler monkeys on Barro Colorado Island in the Panama Canal Zone. He provided normative data on diurnal patterning of activity, social behavior, and vocal exchanges between groups.

An integrated program of studies of social behavior in rhesus monkeys was a highlight of this period. Bernstein and Mason (1963a, b) established a group of rhesus monkeys in a large outdoor enclosure. The intent was to simulate the social structure of a small, natural group in India. They studied basic activity patterns and thus provided basic normative data to serve as the foundation for future studies and for comparison with other species. Studying group formation in the rhesus social group, Bernstein and Mason (1963b) found that clearly demarcated differences in status among individuals and sub-groups were immediately apparent. The monkeys appeared to base these differences on perceived physical attributes, posture, and bearing, rather than in contests of strength.

Sensory/Perceptual Function

A number of studies dealt with sensory-perceptual function. Menzel was interested in the role of perceptual capacity, in addition to motivational factors, in the selection by chimpanzees of the larger of two food stimuli. Menzel (1960, 1961) found that chimpanzees could make choices with the size ratios as small as 1.11 to 1.00. The animals appeared to use mechanisms quite similar to those of humans in making perceptual judgments. Riopelle, Rahm, Itoigawa, and Draper (1964) studied the discrimination of mirror image patterns by rhesus monkeys. Lateral mirror images were more readily discriminated than were stimuli in various other orientations. They concluded that monkeys tend to inspect the bottom edges of the stimuli.

Development

The most active field during this period concerned the development of behavior. During this period much of the program was the responsibility of Mason, Davenport, Menzel, and Rogers. Research in this area can be divided into studies of the effects of early experience, essentially a continuation of the infant studies program initiated by Henry Nissen, and other developmental research.

THE INFANT STUDIES PROGRAM

Work continued on the effects of various rearing environments on the development of chimpanzees. In 1959 a significant addition to the program was made when three African-born chimpanzees were added to the long-term study. First results suggested that these animals were the most active and most fearful of the groups under study. Two years later Riopelle reported the appearance of "stereotyped autistic activities" (e.g., rocking and swaying) in all restricted animals and no wild-born, enriched animals.

Davenport, Menzel, and Rogers (1961) contrasted the lack of effect of "institutional" rearing on mortality in chimpanzees with the results from humans. The chimpanzees isolated from their mothers were actually heavier than those normally reared. The failure to find physical disorders resulting from isolation was regarded as remarkable in view of the studies in humans published during the prior twenty years.

Davenport and Menzel (1963) reported data showing patterns of repetitive, stereotyped movements unique to animals reared in unusual environments. Menzel, Davenport, and Rogers (1963a) studied responses of the various groups to novel objects. The restricted animals were timid with objects and rarely touched them. Restricted animals raised in pairs were the most timid of all.

Menzel, Davenport, and Rogers (1963b) compared responses to novel situations in the animals captured in the field with those reared in small cubicles. The wild-reared animals quickly engaged in a variety of behavioral patterns whereas the isolation-reared animals typically lay still or showed stereotyped behavior.

At this point in the development of the project, the results were still somewhat fragmentary. However, the project would continue well into the years in Atlanta and the results would later be synthesized into a more coherent pattern (e.g., Davenport and Rogers 1970; Mason, Davenport, and Menzel 1968). Perhaps the most remarkable finding to emerge from the program, other than the differences from results with humans, was that chimpanzees were generally somewhat less affected by social deprivation than are rhesus monkeys.

OTHER DEVELOPMENTAL STUDIES

A few representative studies will be reviewed. Menzel (1964) was interested in responsiveness to object-movement. He found that avoidance by chimpanzees increased and approaches decreased with increases in the amount of movement (rate and extensity) of an object. Object novelty and size also were factors. Menzel could use some amounts of movement as rewards and others as punishment in discrimination learning situations.

Mason and Berkson (1962) were interested in the role of tactile stimulation, usually provided by the mother, on the responsiveness of infants to various stimuli. They found that tactile stimulation reduces the arousal level of the infant and thus produces reduced distress vocalizations. This was in an era when arousal theory was popular as a basis for learning and motivation and various phenomena were interpreted in the context of arousal theory.

AGING

A full program of developmental research should include not only studies of young animals but be aimed at the entire life span. Riopelle and Rogers (1965) presented life-span analyses of a variety of physical and behavioral characteristics. They found that with age there were deficits in performance in delayed-response and oddity problems. The older chimpanzees showed improvement in consecutive discrimination tasks. The pattern of deficits suggested a decline with age in the functioning of the prefrontal lobe system of the brain.

Bernstein (1961) compared response variability and rigidity in chimpanzees between eleven and nineteen years of age with those that were between twenty-eight and forty. He found no evidence of a decline in performance and memory or of increases in perseveration, or rigidity with age.

Mental Retardation and Stereotyped Behavior

Prior to going to Orange Park, Gershon Berkson had been studying problems of developmental disorders in humans. He brought that interest to the laboratories and spearheaded a program. One of the major effects of rearing chimpanzees in isolation was the appearance of stereotypies. The study of stereotypy and mental deficiency became fused into a program involving several of the scientific staff members. Work was conducted with both humans and chimpanzees. The work on humans was done at the Sunland Training Center (now Tachachale) in Gainesville. This research program is still regarded as important pioneering work in the field.

CHIMPANZEES

Berkson, Mason, and Saxon (1963) studied stereotypies in laboratory-reared chimpanzees as a function of test conditions. Stereotypies were most frequent in small cubicles. Size per se, was not critical, as there were fewer stereotyped behaviors in smaller outdoor cages than in the cubicles. The availability of objects to manipulate decreased the incidence of stereotypy whereas fear-inducing stimuli increased it. Berkson and Mason (1964a) showed that body rocking and swaying were increased as a function of white-noise level and food deprivation. They were decreased during habituation to a novel stimulus and increased by amphetamine. The data were interpreted as suggesting that increased rocking and swaying were associated with increased arousal.

HUMANS

The Orange Park staff conducted a series of studies of stereotypies in humans with developmental disabilities. Stereotyped movements were found in two-thirds of a group of seventy-one severely retarded patients (Berkson and Davenport 1962). "Stereotyped behaviors were related significantly to self manipulation, IQ, age, and length of institutionalization" (832). Blind individuals displayed higher frequencies than sighted subjects.

Subjects who engaged in more stereotyped behavior manipulated novel objects less than did individuals who showed fewer stereotypies (Davenport and Berkson 1963). There was an increase in stereotyped behaviors in a novel, restricted environment as compared with a familiar environment with opportunities for alternative activities (Berkson and Mason 1963). Giving toys to subjects showing frequent stereotypy reduced the stereotypy and increased manipulation (Berkson and Mason 1964b).

Schusterman (1964b) studied choice of two alternatives, each of which was correct 50 percent of the time, in groups of normal and retarded individuals. He found that subjects attaining a mental age of at least ten years were likely to make choices based on outcomes and responses extending back more than a single trial whereas those with mental ages of five years or less tended to make choices based only on the preceding trial.

Novelty

During this period there was great interest in experimental psychology in the responses of animals to a variety of stimuli that were novel. Much of learning theory had been built on the assumption that basic drives (e.g., hunger, thirst, sexual motivation) and conditioned motives based on association with drive-reducing experiences, controlled behavior. During this time, however, it was found that animals would work in order to be exposed to interesting stimuli, novel stimuli, comforting stimuli, and a variety of situations that appeared unrelated to primary drives. These results suggested major revisions in how complex motivational systems in animals, and especially in humans, might develop. The staff in Orange Park participated in this surge of interest.

Schusterman (1964a) studied responsiveness to a manipulatory object (a piece of rope 10 inches or 36 inches long) in rhesus monkeys in a social situation. He found that the opportunity to manipulate the rope acted as an incentive similar to food.

Menzel and Draper (1965a) examined the effect of size and shape of an aperture on the tendency of rhesus monkeys confined in a small box to look out through peep holes. The larger the effective area of the visual field, the greater was the tendency to look out.

Learning

The long YLPB tradition of studies of learning continued during the Riopelle years. This was a time of wide interest in such studies in experimental psychology.

In oddity discrimination learning problems, animals learn to choose the odd stimulus as correct; available stimuli change from trial to trial. Davenport and Menzel (1960) examined preferences for odd stimuli when no reward is given. Animals were free to choose any of the three stimuli and there were no reward contingencies. Chimpanzees still chose the odd stimulus significantly more than one-third of the time. This suggests the need for some caution in interpreting oddity learning as strictly a function of the reward contingencies.

Schusterman (1962) studied transfer effects in successive discrimination reversal training in chimpanzees. He was interested in the phenomenon of learning to learn in reversal learning. He found that chimpanzees trained on a series of reversal problems, in which the same two stimuli were used but the correct alternative changed in successive problems, transferred almost perfectly to two additional tasks. He suggested that any problem that can lead to a win-stay, lose shift strategy can produce learning sets.

Gibbons had a reputation as being difficult to study in learning situations because they tended to be timid and difficult to motivate. Nevertheless, two studies of learning in gibbons were published. Whereas Berkson (1962) found them difficult when testing them indoors in cages, he had success when testing them outside in view of their cage mates. In studies of delayed-response learning, he found that the gibbons could find the correct location of an incentive in a greater number of trials at delay intervals of 0, 5, 10, and 20 seconds for a preferred food than for a non-preferred food. In single alternation problems the animal must learn to alternate responses (e.g., LRLRLR) in order to get rewarded on each trial. In double alternation problems, the animal must learn to repeat and alternate choices on successive trials (e.g., LLRRLLRR). Schusterman and Bernstein (1962) found that position preferences interfered with learning by gibbons during the early stages and a tendency to alternate responses, independent of the location of the previous reward, interfered later.

The Radiation Program

The radiation program continued, stimulated by concerns about possible nuclear war and the effects of radiation exposure—major social issues of the time. Riopelle (1962c) reported results of the irradiation program that had been conducted by Nissen, Rogers, and himself with both chimpanzees and cynomologus monkeys. There were few concrete results from the chimpanzee work. Nine animals were given 375 or 400 r of whole-body gamma radiation after being trucked to Oak Ridge. Radiation sickness was apparent. Riopelle presented a table of thirteen tasks in which there was no performance decrement and twelve tasks in which some kind of decrement was observed. Many of the disruptions appeared to have been due to effects on motivation rather than to impaired cognitive ability. Riopelle concluded that the irradiation "produced few permanent behavioral losses" (724).

Results with the monkeys were more concrete. Four animals received 2,000 r to the head at the Emory University teletherapy unit, a dose that would have killed them if administered to the whole body. Death occurred 13-35 weeks after treatment. There were various motor deficits shortly before death. There were no deficits in performance in conditioned avoidance tasks that had been learned before irradiation. However, irradiated animals were inferior to controls in visual discrimination learning tasks when stimuli differed with respect to form, color, and size. Riopelle believed that the results could not be attributed to impaired vision or motivation. He likened the effects to those obtained with damage to the posterior association areas of the brain.

Brain Function

The planned program on brain function never developed. The only relevant study published during this period as that of Warren and Akert (1960) in which they showed some deficits in problem solving in cats with lesions of the thalamus. It is not clear whether these studies were conducted in Orange Park or elsewhere. The emphasis on studies of brain function had changed drastically from the Lashley era.

Physical Anthropology

A variety of studies can be classified as dealing with physical anthropology, broadly defined. Some of these blend into the biomedical field. Included was an assortment of studies of blood fractions, growth, dentition, genetics of taste sensitivity, and hepatitis. With the advent of starch-gel electrophoresis, there was growing interest in the genetics of blood serum components. These were useful, for example, in determining paternity before more sophisticated methods were developed. The YLPB scientists analyzed the phenotypes of some individuals in the Orange Park colony and offered a tentative hypothesis regarding patterns of inheritance (e.g., Goodman and Riopelle 1963).

Riopelle (1963a) compared the growth curves for male and female chimpanzees to those of humans. He showed that the weights of humans and chimpanzees were remarkably similar with humans slightly heavier until about fourteen years of age, when the chimpanzees grew much heavier. He used a number of measures, including the age at tooth eruption and closure of the epiphyses to show that the rate of growth in chimpanzees is the faster; they require about two-thirds of the time required by humans to reach the same criteria.

Riopelle (1963b) made a case for chimpanzees as model animals for research on hepatitis. He was making an effort at converting a problem, the spread of hepatitis in the laboratory, into an asset for federal research funding.

Research During the Riopelle Years

As can be seen from the above discussion, the Orange Park scientific staff developed several productive research programs during this period with a strong emphasis on behavioral studies. Although few of these studies might be regarded as groundbreaking, Riopelle had assembled a group that could obtain federal funding, conduct research, and carry it though to publications that were meaningful in the context of the issues of the day. Several of these programs continued into the Bourne era in Orange Park and Atlanta. One can only speculate as to how the Orange Park program might have developed had Riopelle been allowed to build on this foundation for a number of years.

Moving Toward a Move

The Regional Primate Research Center Program

The impetus for the NIH to establish the Regional Primate Research Center system came from a perceived need of the biomedical community for nonhuman primates for research (Dukelow 1995; Yeager 1968). Because these animals are not native to the United States, supplies had been a problem for quite a while.

During the early 1950s there was a series of initiatives and conferences dealing with issues of supply. These led to the formation of several committees to deal with the problem. Because he was considering establishing a breeding program, James Watt, then head of the National Heart Institute (NHI), visited several facilities, including the Russian primate station at Sukhomi, in 1956. He was impressed with the facility. In the Cold War era there was concern that the Russians were maintaining a facility unmatched by the United States. It was the NHI, working with other institutes, that spearheaded the effort to build an American program.

There was then an extended series of political and substantial machinations (Dukelow 1995; Yeager 1968). There was disagreement concerning whether there should be a series of centers or a single large center. Should this be strictly concerned with cardiac function of have a broader mandate? When congressional leaders made it clear that a regional program would be easier to get approved than a single facility, that was the kind of program that was proposed. In 1960 the House and Senate Appropriations Committees allotted $2 million to the NHI to establish two primate centers. An additional $7 million was soon approved. An NHI organizing committee visited thirteen sites. The Orange Park visit by J. Franklin Yeager of the NHI occurred in August, 1959.

The Oregon Center was the first one to gain final approval followed by Washington and Wisconsin. When established in 1961, Yerkes was the fourth center approved. By March, 1962, seven centers had been approved: in Oregon, Washington, Wisconsin, New England, Louisiana, and California in addition to the Yerkes Laboratories. By 1975, the primate Research Centers Program could boast of colonies of 7,800 primates representing forty-eight species, and 141 scientists (Goodwin and Augustine 1975).

Deliberations Between Atlanta and Orange Park

When Riopelle took up residence in Orange Park on June 1, 1959 he was immediately hit with one of the most difficult issues in the history of the laboratories: the effect of the new regional primate research center program on the YLPB. He had to spend the first month getting grant proposals ready for the July 1 NIH deadline. In August he was faced with Yeager's visit and he submitted the formal statement of interest in a primate center.

Riopelle met with a group of Emory officials in October. They were adamant and unanimous in the view that the laboratories should be moved to Atlanta in order to benefit Emory. In a letter to Carmichael, Riopelle added that "a month or two ago, when I was in Atlanta meeting with some of these same people but without visions of sugar plums in their eyes, there was no talk whatsoever about incorporating the Laboratories into the university system or of moving thcm. Indeed, they emphasized that the burden of the Laboratories was mine. Now that there is a possibility of adequate finances, the attitude has changed drastically" (Riopelle 1959a).

Relations between Riopelle and Emory were worsening. Bourne (1959) wrote a scathing letter to Carmichael. He called Riopelle's letters "insulting to the university" and asserted that they expressed the exact opposite of what had occurred. Bourne suggested that the Emory officials had long wanted to move the facility. Bourne wrote that there was a feeling that "Dr. Riopelle is himself too isolated from the Emory environment." He added that "I believe that Dr. Riopelle has shown himself too immature emotionally to control and administer the large grant which he desires and one should consider whether he is, in fact, a suitable person to remain as Director of the Yerkes Laboratories." It may be appropriate to read this letter in light of the fact that it was Bourne who was named to succeed Riopelle.

Board members were concerned. Corner wrote to Carmichael that "we have to deal with an ill-advised president, with two faculties contending for control of Yerkes, and a headstrong, inexperienced director over weak and conflicting control" (Corner 1959). In anticipation of a special meeting of the Board, Richardson (1959a) wrote to the Board members explaining Emory's reasons for wanting to move the facility to reduce its isolation and facilitate its expansion. Richardson enclosed a memorandum in which Lashley (1958c) favored moving the facility. Lashley was especially concerned that the town of Orange Park was beginning to develop at such a rate as to suggest serious infringement on the site. He argued that the climate was not really ideal for chimpanzees, the staff was isolated from colleagues, costs of labor and housing were increasing, the colony was too large, and the importance of chimpanzees as research animals had been exaggerated. Some of Lashley's views had not changed in twenty years.

Riopelle met again with Emory officials on November 13. No signs of appreciable progress were apparent. At one point, "Dean Lester raised the question of where Dr. Riopelle stands, with us or against us" (Minutes November 13, 1959 meeting). The widening gap between sides was clear.

The special meeting of the Board, along with some Emory officials, was held at the Smithsonian Institution in Washington, DC on November 25. Bourne and Richardson presented a plan for a center of which the laboratories would be a part. Carmichael argued strongly that an application should be made for the YLPB as an expanded facility rather than as part of a larger center. Bourne and Richardson got tougher, noting that the YLPB would not be allowed to expand at its present location. Emory President Martin was frank in noting that the possibility of obtaining an NHI grant precipitated the issue of moving the laboratories to the Emory campus but urged a united front. Richardson exclaimed "We at Emory must have access to primates and we will have them!" (Board minutes November 1959, 6).

Somehow, Carmichael sensed agreement. He summarized the consensus that "the new facilities are to be built in Georgia but that the chimpanzee facilities will remain for at least some time in Orange Park. At the proper [time] in future years they can be moved to the Georgia site when that site has been developed and is in full scientific operation" (Board minutes November, 1959, 6). That proved unrealistic.

Emory's grant application to the NIH was dated December 10, 1959 with Richardson as the Principal Investigator on behalf of a proposed "United States Public Health Primate Center (Emory University)" (Richardson 1959b). The chain of command would flow upward from the Director of the Yerkes Labs to a Director of the Primate Center to the President's Primate Center Committee. The proposal called for a continuation of behavioral studies but with a substantial increase of more biomedical research. The cost of the proposed buildings was $3.8 million.

With the Emory proposal already on the table, members of the NIH National Committee on Primates, Karl F. Meyer, a Mr. Audy, Willard H.

Eyestone, F. Schemehl, and Katherine Parent, together with Richardson, visited Orange Park on January 12, 1960 (Dukelow 1995). Riopelle (1960a) reported that things went well. The visitors favored having the facility remain in Orange Park. He wrote to Carmichael that members of the committee urged him to submit an application for basic support of the laboratories much as he had to the NSF.

An NSF team visited soon thereafter. It was from the team that Riopelle learned that Emory was trying to move the chimpanzees to Atlanta as soon as June, 1961. Even more distressing to Riopelle was the fact that Emory failed to share these plans with the director; he had to learn of them from the NSF site visit team (Riopelle 1960b).

In between the two site visits, Parent, of the NIH committee, had a telephone conversation with the influential Boisfeuillet Jones (Dukelow 1995), then an Emory vice president but soon to move to Washington in the Kennedy administration. Jones lobbied for the move to Atlanta. In the words of Dukelow (1995), "it is evident, to anyone reading the historical accounts and correspondence, that intrigue and unstated facts were clouding the application for primate center status" (118).

The first NHI support funds were already in place when the board next met, in Orange Park in June 1960. Disagreement about primate center status persisted. Bourne and Richardson indicated that the final proposal was to move the facility to Atlanta in about two years. It was Carmichael's understanding that the grant was to support the facility in Orange Park. He was concerned about disrupting ongoing research and that it would take longer to establish the Emory facility as an ongoing scientific institution. Carmichael expressed other concerns related to the ability of the staff to work full-time on research if they were on a university campus and lacking the isolation from urban pressures that a more rural location could provide.

On December 7 Jones intervened again, this time telephoning Eyestone asking that Riopelle's name not be placed as the principal investigator on the primate center application but that Jones' name replace it. Jones expressed concern that Riopelle tended to act on his own. The change may also have been because Riopelle's name was already becoming linked to the Tulane University application for a center (Dukelow 1995).

Riopelle did not give up. He tried to do an end run around the Emory officials by working with the board (e.g., Riopelle 1961). The position was complicated because by this time Jones, now an Assistant Secretary of Health, Education, and Welfare in the Washington, was again lobbying for the Emory position. Jones apparently had oversight authority for the

NIH and was also a candidate for the position of president of Emory (Mullen 1962; Riopelle 1962a).

Riopelle's position soon worsened. That same month, Orie Myers informed Eyestone of Emory's search for a new director for the Center. On January 12, Myers notified Eyestone that Bourne had been chosen (Dukelow 1995).

A February 2, 1962 memorandum outlined the program for the planned Center. Bourne would be the director with W. C. Osman-Hill, Riopelle, and ten others from various departments as associate directors. There would be a dramatic shift toward biomedical research with programs in gross anatomy, embryology, neurology, endocrinology, chromosome studies, nutritional studies, cardiovascular studies, pharmacology, experimental surgery, pathology and microbiology, immunology, biochemical studies, histochemical studies, and fine structure. The Emory proposal seemed to include the entire medical school. All animals would be moved to Atlanta. On March 2, 1962 a group of Emory faculty and officials met with Eyestone, Richard Hensch, and Yeager from the National Heart Institute to discuss details of the new primate center. (Summary of meeting of March 2, 1962).

The public announcements of the new primate center grant and the move to Atlanta were not handled well. Rumors of a possible move circulated in the Jacksonville area and the public statements were not always completely forthright.

The minutes of the March 26 and 27, 1962 Board meeting concluded with the following announcement: "Moments before adjournment the Florida Times Union of Jacksonville telephoned the Laboratories that National Institutes of Health had released news that Emory University had been granted the award for Regional Primate Center" (Board minutes March, 1962). Hisaw (1962b) later wrote that "Emory had engaged in considerable political maneuvering in support of their proposal."

For at least the short run, it appeared as though a multi-headed monster would be running the facilities. The situation was virtually guaranteed to lead to discord with the Board, Riopelle, and Bourne treading on each other's toes with divided lines of authority. To take but one example, Bourne hired anthropologist W. C. Osman- Hill, an individual who had previously been considered and rejected for the post as director. Hisaw (1962a) was concerned about Bourne's authority to make appointments and for the future role of the Board. It was unclear whether the appointment was for the YLPB or Emory and where the dividing line lay.

Another special Board meeting was held in Orange Park on December 10, 1962. This meeting was mainly to tie up some loose ends. With Carmichael unable to attend, Hisaw (1962c) reported that he "came away feeling as though I had presided at the funeral of an admired friend." Riopelle formally relinquished the directorship of the YLPB on December 31; Bourne took over the next day.

Community Involvement

The maneuvering was not confined to those with a direct stake in the laboratories. Interests in the Jacksonville area became involved while this was transpiring. Public awareness of the proposed move arose in February 1962 when two local physicians met with Riopelle and he explained why he could not make long-term commitments (Mullen 1962). A group of Jacksonville citizens then met with Riopelle to discuss the situation. Telegrams protesting a move were sent to both the NHI and Emory. The media were then brought in. Walter G. Odum, mayor of Orange Park got involved, citing the economic impact of the facility.

Congressmen Charles E. Bennett and D. R. Matthews soon entered the discussion and tried to wield influence in Washington. They argued that Orange Park was already an economically depressed area because of the recent closing of a nearby naval base and that moving the facility would worsen the situation. Other local officials and civic leaders pitched in (Mullen 1962). Resolutions opposing the move were presented by the Duval County Medical Society, the Jacksonville Area Chamber of Commerce, and the Jacksonville Alumni Association of Emory University.

The possibility that Jacksonville University, a relatively new private university, might assume sponsorship was proposed and their officials were receptive to such a proposal (Mullen 1962); Emory was not (*Florida Times-Union* 1962b). Harold Colee, executive vice president of the Florida State Chamber of Commerce, also issued a statement (*Florida Times-Union* 1962d).

Jacksonville's *Florida Times-Union* followed developments with a series of stories on the unfolding saga. On March 19, a week before the Board meeting in Orange Park, the newspaper ran a detailed story on the history and activities of the YLPB (Drane 1962) that appeared intended to generate local interest in the YLPB.

According to Mullen, Myers conducted a news conference at the conclusion of the March board meeting and indicated that the decision to move the laboratories was contingent upon receipt of NIH funds even though he had already been informed during the previous week that funds had been granted (Mullen 1962). The newspaper confirmed Mullen's version (*Florida Times-Union,* 1962c). At that point, Myers's public stance was that the decision as to whether to move the YLPB to Atlanta or keep it in Orange Park had not yet been finalized; of course, it had been.

Florida University Involvement

Yet another sub-plot progressed in parallel with these events. This concerned possible support for the YLPB from the University of Florida or other institutions within the state system. Several individual University of Florida faculty members had developed cooperative research projects with individuals on the YLPB staff but there was no formal affiliation. In October 1959 Riopelle journeyed to Gainesville to discuss the possibility of University of Florida involvement in sponsoring the national primate center (Riopelle 1959). The Florida medical school was brand new, having just opened in 1957, and not yet in a position to enter into such a venture.

University involvement was revisited in 1962 when Riopelle again tried to enlist the aid of the University of Florida in preserving the facility. Gert H. W. Schmidt, of the Board of Control of Florida, the body representing the whole system of Florida universities, visited Orange Park at Riopelle's request and became an enthusiastic supporter of university involvement. In an April 6 memorandum, Schmidt (1962) argued for a purchase of the facility. A committee was formed to study the possibility but no action was taken. On his way out the door in December, Riopelle (1962b) wrote to President J. Wayne Reitz of the University of Florida strongly urging him to move forward. Reitz (1962) had studied the possibility and concluded that the purchase was not consistent with the mission of the university at that time. In 1964 Emory would offer the facility to Reitz for $250,000; the offer was refused.

Dean George T. Harrell, Jr., of the University of Florida Medical School noted that he had disagreed with Reitz and had an interest in taking over the facility and moving it to Gainesville (Harrell 1980, 249). The university eventually filed a separate application for a primate center; it was declined. However, the NIH did provide funds to construct a smaller primate facility (Harrell 1964).

Perspective

Obviously, while all of this was going on, the staff members were concerned about the uncertainty of their futures. Most thought that the

YLPB should remain in Florida (Draper 1994; Mason 1994; Rogers 1994). Berkson (1993) recalled that "Riopelle would get us together and tell us in excruciating detail the latest developments." Later he added that "apparently he was sharing his anxiety with us, which of course made us terribly anxious about what was going to happen . . . so I always thought of it as psychotherapy for him but it was a little bit burdensome on us" (Berkson 1996)

Clearly, Carmichael, Riopelle, and some members of the board believed that Robert Yerkes had made a good decision in locating the facility in Orange Park and that a national center should be located there. On the other side, the personnel at Emory had had the valuable facility fall into their laps and believed that they could best serve their institution by consolidating it with other programs and facilities on their campus. It was perhaps because both positions were reasonable ones that there was such conflict and such maneuvering. Obviously, the climate of Orange Park, although not ideal, is more amenable to outdoor housing of chimpanzees that is that of Atlanta. On the other hand, Orange Park had grown to such an extent that it is a very different place than the one that Robert Yerkes chose and further development would have infringed almost as much on the facility in Orange Park as in Atlanta. The university ties and urban environment surely were an attraction in Atlanta. In essence, one can see advantages to both positions.

Much later, Riopelle (1994) noted that his views on the move had shifted back and forth over the years. He still believed that Orange Park would have been the right place had it stayed as it had been. With time, he clearly saw both sides in a different, more balanced, perspective. As always, hindsight was beneficial.

15

The Bourne Years and the Transition to Atlanta

GEOFFREY BOURNE ASSUMED THE DIRECTORSHIP OF THE LABORATOries, by then the Yerkes Regional Primate Research Center (YRPRC), on January 1, 1963. The laboratories had 2½ years remaining in Orange Park under his directorship. The primary focus during this period was upon the transition to Atlanta. Bourne continued to reside in Atlanta but often spent several days per week in Orange Park. He ran the new YRPRC as a two-pronged effort with one base in each location. Because this is a book primarily about the Florida facility and because much of the research conducted in Atlanta was conceptually distinct from the Florida effort, I will concentrate on the work in Orange Park.

Writing about this period is made difficult because there is less information available than for previous periods. The 33rd and last of the traditional annual reports was prepared by Riopelle in June 1961. The report of June 1962 was prepared by Murray Townsend and included only a financial statement. Bourne did prepare undated annual reports for 1963–64 and 1965–66. The relevant archival folders suggest that no report was available for 1964–65. I have located just two sets of minutes from meetings of the Board of Scientific Advisors, those of October 1963 and November 1964. Although these reports provide much valuable information, they are less inclusive that those in the older format.

There was relatively little innovation during this period of transition. The focus was rather upon completing research in progress, sorting out the futures for both human and nonhuman primates, and preparing for a major reorganization. Perhaps the most notable exception was the flurry of field studies of primate behavior that were conducted during this period.

Photograph of Geoffrey Bourne, fifth director. From the Yerkes National Primate Research Center, Emory University, Atlanta, GA.

The Death of Ada Watterson Yerkes

Ada Watterson Yerkes died at her home in New Haven on October 16, 1963—less than a year into the Bourne administration's tenure. Her death can be viewed as symbolic as the loss of a last link to the establishment of the Orange Park facility. The times were changing and the older generation was disappearing.

The New Director: Geoffrey Bourne

The always colorful Geoffrey Howard Bourne was born in Perth, Western Australia on November 17, 1909 (Dukelow 1995). Bourne was first married in England. The couple had two sons. After a difficult divorce from his first wife, Bourne married Maria Nelly Golarz, PhD in 1964. The two collaborated on many research projects.

Bourne received a Bachelor of Science degree in 1930, a Master of Science degree in 1932, and a Doctor of Science degree with a thesis on the histology of the adrenal gland in 1935, all from the University of Western Australia. In addition, he received a Doctor of Philosophy degree from the University of Oxford in 1943. From 1931 until 1938 he held a variety of positions in Australia, finishing as a biochemist for the Commonwealth of Australia Advisory Council on Nutrition in 1938. Bourne lived in England during 1938 through 1957. He was credited with first demonstrating the presence of Vitamin C in cells.

Bourne served as a Reader in Histology at the University of London during 1947–57, when he assumed the chairmanship of the department of anatomy at Emory University. He is credited with having major impact on a department that had been rumored to be about to lose its accreditation. Bourne secured a large NIH training grant and brought in new faculty and students to bolster the program.

Bourne authored or edited many books; Dukelow (1995) credits him with nearly 100. Among those that he authored are *The Ape People* (Bourne 1971) and *Primate Odyssey* (Bourne 1974). He was the author or co-author of more than 600 scientific papers.

In 1978 Bourne retired from Emory and became vice chancellor and professor of nutrition at St. George's University School of Medicine on the island of Grenada. In 1983 a coup organized by a revolutionary council that deposed the prime minister of Grenada prompted U. S. President Ronald Reagan to call for an invasion to rescue the students, many of them American, at the university (*New York Times,* 1988; Golarz de

Bourne and King 1990). Bourne mediated among the political, diplomatic, and military forces of Grenada and the United States in an effort to avoid conflict and spare the medical students.

Dukelow (1995) highlighted two of Bourne's characteristics that he thought especially outstanding. One was his extraordinary productivity as an author. According to Dukelow, "he was not a perfectionist in his writing, arguing that if he had to write things over and over there would be many things he could not do" (122). The second characteristic was a love of public relations. Bourne loved to associate with important people—actors, royal families, and others. He was a showman. He appeared on the Johnny Carson television program demonstrating how he taught chimpanzees to communicate by typing on a computer and was photographed with various dignataries. He died of heart failure in 1988.

In reading documents and interviewing YRPRC personnel, one sometimes get the impression that Bourne gained somewhat less respect than some of his predecessors. There were questions about the way in which he gained his appointment. In his administration, he tended to be easily swayed by those who took strong positions. There was a feeling that he had no core research program or direction. Rather, he brought in scientists from other countries who published articles that included Bourne as a co-author. Although he had a high profile, some staff members were less confident of the substance of his scientific work. He wrote a lot but it often appeared superficial. However, he was sociable and a very cordial host.

Administration

At the time of the time of the 1963 Board meeting, Long remained as chair with Richardson as associate chair. The remaining members were Carmichael, Bernard Holland, Harry F. Harlow, Frederick J. Stare, Taliaferro, and Bourne, member *ex officio*. Stare was a Harvard biochemist specializing in studies of nutrition. With two vacancies on the Board due to resignations from Corner and Hisaw, the Board voted to add Richard T. Smith, and Knut Schmidt-Nielsen to membership. The latter apparently declined the invitation. One year later, Helmut O. Hofer and Niko Tinbergen had replaced Schmidt-Nielsen and Stare. Hofer was a Viennese anatomist who moved from Europe to the Delta Regional Primate Research Center in 1965. There is no indication that Tinbergen, the Nobel-Prize winning Oxford ethologist, ever attended a meeting. YLPB alumnus C. R. Carpenter joined the board shortly after the period covered here.

Personnel

Scientific Staff

Perhaps the most notable aspect of the personnel list during the Bourne years in Orange Park is the relative absence of new blood. The focus was upon completing projects in progress in order to prepare for moves from the site. Bernstein, Davenport, Draper, Fitz-Gerald, Menzel, and Rogers remained on the staff and were committed to move with the facility to Atlanta. Riopelle had left in December 1962; Mason moved to Tulane with Riopelle after a year of field work in Colombia. Berkson moved to Chicago and Schusterman to California.

Jan Moor-Jankowski, a Polish MD with an interest in immunogenetics, began working in Orange Park as a visitor in late 1962 or early 1963 and joined the staff for one year in 1964. After leaving the Yerkes laboratories, he had a long career at New York University.

Veterinarian Norman B. Guilloud joined the staff in Orange Park in July of 1964, the first full-time veterinarian to be hired in Orange Park. He had received his education at Texas A & M University and gained experience working with primates while in the U.S. Army at Walter Reed Hospital. He later moved with the center to Atlanta.

The key appointment in Atlanta was that of William C. Osman-Hill as associate director. Although based in Atlanta, he also spent some time in Orange Park. Osman-Hill was an anatomist with an MD degree from the University of Birmingham. He was remembered by some staff members as the archetypical English scholar-gentleman who viewed those from the "colonies" as a step below the British.

Technical and Service Staff

Many members of the secretarial staff agreed to move to Atlanta; the exceptions remained because of marital commitments. The Townsends, key personnel, would move to Atlanta, along with their son Roger, who also worked there and was an experienced animal caretaker. Murray Townsend, who gained greater influence with the director off site, is reported to have often sided with Bourne and favored his efforts in contrast to those of Riopelle. She seemed disturbed by those staff members who decided to go to the Delta Primate Center.

Two other notable staff members moving to Atlanta were Judy and Jimmy Roberts. Jimmy would succeed Graham Townsend as the colony superintendent in 1969 and Judy then would be the bookkeeper. Judy

Roberts began working at the YLPB in 1959, soon after Riopelle arrived (Judy Roberts 1993). She started as a Dictaphone typist and was a technician working on the developmental research with Davenport, Mason, Menzel, and Rogers. Jimmy began work as a caretaker in 1963 fresh off the farm at the age of twenty-three. He recalled many nights staying up with apes in labor.

Judy Roberts (1993) recalled a trip when Bourne was to appear with an orangutan on the Dick Cavett television show. Because the ape could not fly, the Roberts were chosen to drive the animal to New York. She recalled a difficult drive in a snowstorm, the problems of keeping an orangutan in a hotel room, and harrowing taxi rides around New York City. She recalled "I know now what smugglers and all feel like" because they "could not go into a restaurant or engage in various other activities common to most travelers."

Louise and Clyde Wright were two other key staff members. Louise started as a laboratory assistant for Davenport and Rogers worked as a technician with Mason. By the time they left Orange Park, she had been promoted to be the secretary to the director. Clyde worked with Bernstein as a technician in Orange Park. Both moved to Georgia with the station. Clyde became the first superintendent at the Field Station of the YRPRC. (Bernstein 1993; Orange Park 1991).

In his Annual Report of 1964, Bourne reported that "We have been fortunate in getting all the colored animal caretakers whom we want to sign letters of agreement that they will come to Emory" (4).

Jesse Graham Townsend had to retire from the YRPRC in April 1969 for health reasons and died in his sleep the night of January 1, 1970. A plaque in Townsend's honor was placed in the center in January 1972. (Many Friends 1972). At the ceremony it was recalled that Lashley had called Townsend "the greatest authority on captive chimpanzees in the world" (26). When I visited the Center during the 1990s nobody was able to locate the plaque.

Personnel Problem

Some difficulties were experienced in the relationship between Director Bourne and Moor-Jankowski. According to Moor-Jankowski (1994), he left the YRPRC after discovering that tuberculosis was widespread in the Orange Park colony and one of his technicians came down with TB pleurisy. When he refused to remain silent about this he was asked to leave. The attending veterinarian, Norman B. Guilloud (2002), disputed this, noting that "I personally tested all non-human primates at Yerkes and found no tuberculosis via tuberculin testing and radiography. No necropsies revealed tuberculosis during the time I was in {Florida] from July, 1964 until we moved to Atlanta in the spring of 1965."

At a minimum, however, there were other issues involved. At the November 1964 board meeting, Bourne reported that he had asked Moor-Jankowski to seek another position. According to Bourne, Moor-Jankowski resisted administrative control, failed to hold expenditures within budget, acted irresponsibly with regard to the health and welfare of the animals, gave forbidden injections, did excessive bleeding, and made untrue statements about the laboratory. Moor-Jankowski asked to appear before the board to present his case. The board decided that it should not serve as a court of appeals but asked Board members Taliaferro and Smith to meet with him and report back. They were impressed with his research and his picture of immunology but noted that interpersonal problems existed. The board supported the director's decision.

Bourne was especially concerned that Moor-Jankowski wanted to conduct a series of cross-immunization studies with orangutans that might interfere with future breeding. By this time the YRPRC had built an important collection of orangutans and he was concerned that invasive research might jeopardize the colony. Guilloud (2002) was concerned that Moor-Jankowski, without permission, "had given Complete Freund's Adjuvant (CFA) to chimps. Since CFA contains Mycobacterium tuberculosis capsule, it causes the body to produce antibody to TB and thus causes animals to show a (+) reaction to tuberculin which means that tuberculin can not be used to test animals for TB. This was a very serious violation of good laboratory practices and should not be done!"

Clearly, there were two sides to the story. In the end, however, Moor-Jankowski left after one year.

Funding

The problems of funding that had plagued the YLPB for so many years had now been greatly eased by primate center status and the accompanying dollars. Bourne could report that "perhaps for the first time in the Laboratories' history, the Advisors need not worry about finances" (Board Minutes 1963). As with other characteristics of the facility during this period, however, information regarding funding is less complete than it had been in previous years.

The YRPRC was awarded a yearly operating grant of $550,000 as part of the regional center allocation. In addition to that, the "heart" grant of $150,000 per year continued. Those totals did not include the costs of construction for the new center; they were estimated at that point as $200,000 for the field station and $1,483,000 for the campus facilities. The university supplied the land but building costs were absorbed 100 percent by federal funds.

These general allocations were supplemented by a series of individual grants. In the Annual Report of 1964, Bourne provided a list of grant accounts but did not provide dollar amounts or any breakdown of the income or expenses. Included were a Ford Foundation grant, the NIH-USPHS operations grant, an NIH grant in nutrition, an NIH grant to Mason, an NSF grant to Bernstein, a contract from the Office of the Surgeon General of the U.S. Army for work on hepatitis, a grant from the National Association for Retarded Children to Berkson, an NIH grant to Menzel, and a supplemental operations grant from the NIH. In this report he listed the two support grants at $172,000 and $481,100, $398,100 of which had been approved as of the writing of the report. In addition, three of Bourne's grants were transferred from the Anatomy Department to the Center. The new era of Federal funding had arrived and budgets mushroomed.

Physical Plant

There is no indication in the materials available that any substantial funds were spent on maintenance or improvement of the Orange Park facility. The focus was on the new center in Atlanta. The main center would occupy twenty-six acres with a main building that contained in excess of 50,000 square feet. The groundbreaking ceremony occurred on December 30, 1963. The field station, to be built near Lawrenceville, twenty-five miles from the campus, would contain office space, a small operating theater, testing rooms, ape units, small primate facilities, and a large compound.

Animals

Bourne expanded both the number of animals and the range of species in the Center. The inventory published in the November 1964 issue of the *Yerkes Newsletter* included nine lowland gorillas, sixty-nine chimpanzees, twenty-seven orangutans, three cebus, twenty-four Celebus black apes, one Diana monkey, four drills, one hybrid macaque, one Java macaque, one mustached guenon, forty pigtailed macaques, one rhesus monkey, seven spider monkeys, ten stumptailed macaques, five vervets, and one wooly monkey.

Osman-Hill was appointed to chair a committee on animal welfare. When Guilloud came on board in 1964 as the first veterinarian, a number of problems soon became apparent and these led to some interesting publications in comparative medicine. As he arrived in Orange Park, Guilloud was called to treat an orangutan and several gorillas who turned out to be the first known cases of paralytic polio in nonhuman primates (e.g., Guilloud and Kline 1966). Another syndrome documented by Guilloud for the first time was hyperthermia in gorillas (Guilloud and Fitz-Gerald 1967).

Other Activities

A welcome innovation initiated by Bourne was publication of a *Yerkes Newsletter,* originally under the editorship of Frances Fitz-Gerald. Much information that might otherwise be presented only in reports with limited circulation was included in the *Newsletter.* This publication was later replaced by a more sophisticated publication, *Inside Yerkes.*

The laboratories continued to provide specimens and samples to other laboratories around the world. In his 1963–64 Annual Report Bourne listed at least twenty-five institutions in the United States and Germany that had received blood fractions.

Life in Orange Park

With all of the changes that were occurring in the laboratories and in society, some aspects of life in Orange Park remained unchanged. Guilloud (1993b) recalled that during his stay in Orange Park it had become very much a Navy town. The street on which he lived was dominated by Navy pilots, at least one of whom was killed while he was there. By then, the town accepted the laboratories with little question. He enjoyed his stay there.

One strong memory reported from this period was of a strong hurricane, presumably either Cleo or Dora in the summer of 1964. Many members of the staff and their families spent a couple of days and at least one night in the laboratory buildings, in part because these buildings were on high ground and more solid than their homes and partly out of loyalty to the animals. In fact, although the laboratory experienced some wind damage and lost power, no cages were destroyed and no animals escaped (Guilloud 1993a).

Research During the Bourne Years

The YRPRC now had two centers of research activity: Atlanta and Orange Park. Most of the work at the former was conceptually different from that at the latter and with the main themes of this book. Therefore, in this section I deal primarily with work that appears to have been done in Orange Park. Because some projects were continued, written up elsewhere, or published well after the research was completed, I have had to be flexible with rules for inclusion.

The first notable change apparent in examining the official bibliography of the YRPRC is a sudden increase in the number of published articles. The total number of publications listed for 1958–62 was seventy-seven (mean of 15.4 per year). For the next five years it was 252 (mean of 50.4 per year) or more than a threefold increase. Most of this was due to the inclusion of the Atlanta work but some of it was also due to the increased activity in Orange Park stimulated by Riopelle and lagging into publication this period because of the delays encountered between the completion of research and appearance in print.

Studies of Basic Behavioral Patterns

Bernstein (1965) studied the activity patterns of a group of twelve cebus monkeys with the same methods and the same apparatus as with earlier studies of rhesus monkeys and gibbons. He found intensive social interactions to be relatively rare and dominance to be relatively unimportant. Each of the species appeared to display its own characteristic pattern of activity. Patterns of marking and chest rubbing were unique to the cebus monkeys among the three species studied.

Whereas adult gibbons, like most primate species, show relatively little sharing of food, Berkson and Schusterman (1964) found that young gibbons often shared. Both the frequency of attempts and the success of the attempts were affected by individual differences, degree of hunger, proximity, and the availability of food. The authors suggested that these patterns may be adaptive under natural conditions where some young animals might have difficulty securing food in the arboreal environment.

Bernstein used the term "control male" to refer to individual males in troops of pigtail monkeys that had a special role in protecting the group from internal and external challenge. Bernstein (1966a) studied responses to challenge in groups in outdoor enclosures. The group was challenged in a variety of ways including by a human who would restrain a group member or by placing an object (e.g., a snake, a turtle, a spider monkey, bananas, or a wooden block) in a box that could be opened by a monkey. The control males were most involved in responding to the challenge, especially when the challenge was direct (e.g., a human). Bernstein (1966b) found similar roles in rhesus, pigtail monkeys, and capuchin monkeys with similar challenge tests. The role of control male was not dependent on dominance interactions. Bernstein suggested that the control males are more important to the organization of primate societies than are dominant individuals.

Mason (1964) presented a wide-ranging review of social behavior in nonhuman primates as it was understood at the time. He emphasized the gregariousness of most primate species. Mason suggested that social organization appears to be more rigid in terrestrial baboons and macaques than in arboreal species. Mason, like other authors, suggested that the selective pressures that led to the evolution of animals capable of solving many complex discrimination learning problems in the laboratory may lie in the need to monitor social interactions in groups in nature.

Field Studies

This was a period of increased interest in field studies of groups of free-ranging nonhuman primates. Such work was envisaged by Robert Yerkes, conducted by Henry Nissen and Harold Bingham, and developed on a substantial scale by C. R. Carpenter. The work of Jane Goodall, George Schaller, and others brought such research to prominence during this period. The Yerkes staff scientists wanted to participate. One of the inducements that Bourne used to get people to agree to move to Atlanta was funding for field research (Bernstein 1993).

Davenport (1967) reported results of his study of orangutans in Malaysia but appeared frustrated with the lack of information regarding such patterns as general social organization, dominance, infant care, and sexual behavior.

Menzel's studied Japanese monkeys in Japan in a population already under intensive study by Japanese scientists. Menzel (1966) focused his study on responsiveness to objects including naturally occurring objects, novel toys, and novel dolls. He found that spontaneous manipulation of objects was generally both less frequent and less persistent in these feral monkeys than in laboratory animals.

Draper (1966) studied age and sex differences in activity patterns in the free-ranging colony of rhesus monkeys originally established by Carpenter on Cayo Santiago, Puerto Rico. He found most activities, such as postural adjustments, locomotor behavior, manipulation, and other nonsocial activities, to decrease as the monkeys age.

Bernstein's main focus in his research in Malaysia was upon pigtail monkeys. He found the study population to be badly disturbed by human intrusion (Bernstein 1967b). He described the travel patterns, feeding, agonistic patterns, sexual behavior, play, and other responses in these animals. The behavioral patterns observed under these conditions were basically the same as those seen in the laboratory, although the frequencies of various patterns differed. Bernstein (1967a) found five different taxa of nonhuman primates, pigtail monkeys, gibbons, irus macaques, and two species of langurs, in a single small forest in Malaysia and studied their interactions. The animals of different species interacted fairly frequently and peacefully.

Bernstein (1967c), in a characteristically reflective article, discussed the very meaning of the concept of a natural habitat. He suggested that because of spatial limitations, temporal factors, human disturbance of the habitat, and human predation, it can be difficult to determine exactly what kind of habitat is truly natural for a given species.

Sensory/Perceptual Function

Menzel extended his interest in perceptual factors in food selection in chimpanzees to orangutans, gibbons, and rhesus (e.g., Menzel and Draper 1965b). Animals could choose between a standard, but invisible banana, and a visible banana of varying size. All animals showed an approximate 2 percent "bird-in-hand" preference for visible fruit, although the species differed with respect to the accuracy of their food size selection. The authors followed up this work with studies of chimpanzees' performance in using differently colored plaques as cues to food rewards of different sizes.

Development

EARLY EXPERIENCE

Much of the research during earlier administrations had been conducted with animals tested relatively soon after and during rearing under varying conditions. The research continued during Bourne's administration in both Orange Park and Atlanta. The animals reared under varying conditions could now be tested as they approached adulthood. Davenport, Menzel, and Rogers continued the program initiated by Henry Nissen. Previous research had shown that infant chimpanzees reared under conditions of extreme environmental restriction remain in better health and weigh more than infants reared by their mothers. This is contrary to findings with humans. However, infants reared without their mothers tended to develop stereotypies, odd and persistent motor patterns not generally seen in mother-reared animals. Davenport, Menzel, and Rogers (1966) took three wild-born infants who had been reared by their mothers for several months and exposed them to six months of extreme restriction at approximately eighteen months of age. The infants remained in good health and did not develop the stereotypies displayed by the other animals. These results suggest an important role of the timing of the restriction.

A series of three studies was designed to examine the effects of differential rearing conditions on intellectual performance in chimpanzees. In one, Davenport and Rogers (1968) compared the performance of chimpanzees separated from their mothers at birth and reared for two years in a restricted environment (the small gray cribs) with that of wild-born animals reared in enriched environments on spatial delayed response tasks at 7-9 years of age. Initially, the restricted animals were inferior to the enriched subjects but, with experience, they performed at a similar level to the control animals. The authors suggested that the early differences were due to the manner in which the restricted animals approached the task rather than to any cognitive difference. In essence, the restricted animals tended to look away from the apparatus during delays, to explore the cage, and to engage in stereotyped movements whereas the enriched animals remained more task oriented. Improvement was shown when these distracting behaviors were reduced. The authors suggested that human children reared in unstimulating environments experience similar problems in IQ tests.

Duane Rumbaugh has long been an advocate of the transfer index in comparative studies of learning. This entails training animals on a discrimination to a given criterion and then conducting a series of reversals in which the previously correct stimulus becomes the incorrect alternative and vice versa. Davenport, Rogers, and Rumbaugh (1973) used the transfer approach and found that the deprived chimpanzees performed at a rate that was inferior to that of wild-born animals. Because all animals learned the initial problems to a given criterion, the authors suggested that the results were due to deficits in cognitive ability and could not be attributed to emotional, motivational, or attentional factors.

Rogers and Davenport (1969a, b) found the sexual behavior of five male and seven female chimpanzees reared in isolation for the first three years of life to be initially deficient. Unlike the famous rhesus monkey studies of Harry Harlow, however, these studies showed that some chimpanzees reared in restricted environments could learn to copulate with sophisticated partners. Rogers and Davenport found that substitution of human interaction during infancy was actually detrimental to the development of sexual competence. They suggested that the difference between rhesus monkeys and chimpanzees lies in the less stereotyped sexual behavior of the latter, which makes it more amenable to later learning.

Mason et al. (1968) provided a useful overview of the early experience research, particularly that related to social behavior, on nonhuman primates. They concluded that experience is essential to the development of normal social behavior in both rhesus monkeys and chimpanzees. However, "the evidence suggests that some effects of social deprivation are more persistent and severe among monkeys" (36). They suggested that this may be because the chimpanzees possess more highly developed perceptual and cognitive abilities that they can use in overcoming the effects of social deprivation.

OTHER DEVELOPMENTAL STUDIES

Rogers and Davenport (1970) tried to synthesize all of the data collected since the founding of the Orange Park facility in 1930 concerning maternal behavior in chimpanzees. They found that the rearing condition of the mother had a profound effect on her maternal behavior. Those mothers who had extensive contact with their own mothers were much better mothers on their own. Further, maternal behavior improved with successive births. They provided a theoretical interpretation of the interaction of the infant and mother that is in the spirit of the arousal theory that was the prevalent in psychology and the YRPRC.

Berkson (1965) studied the patterns of eye fixation in response to various stimuli in infant chimpanzees. He found that eye fixation declined during the first minute of presentation. Presentation of two stimuli slowed the course of habituation to the stimulus. This is essentially a study of attention, although the word would be used sparingly in 1965. Similar methods would be later used with young human children. In this respect, this was a pioneering study.

Mason (1965) tried to synthesize many data on the determinants of social behavior in young chimpanzees. He proposed that a variety of different procedures increased the general state of excitement, or arousal, of young chimpanzees and that these, in turn increased the social behavior of the animals. Intense auditory stimulation, strange surroundings, separation from a cagemate, physical restraint, and novel social stimuli all affected social behavior. Social interaction appeared to lower arousal levels. Mason suggested that this model could help explain the nonspecificity of effect (i.e., that many different variables produced similar effects). Changes in arousal level appeared to be reinforcing and this might suggest why social interactions also appeared to have reinforcing properties.

Mental Retardation and Stereotyped Behavior

Berkson continued his interest in mental retardation in humans while also working with chimpanzees. In a review article, he (Berkson 1963) linked some differences between normal and retarded human subjects in psychophysiological responsivity to differences in arousability in a manner similar to Mason's use of the concept. Berkson proposed that "mental defectives are less responsive to stimuli of short duration than are normals, and there is reason to believe that such a defect in responsiveness is related to the diffuse brain abnormality found in this group" (569).

Learning

The interest in the study of learning that characterized the earlier years of the YLPB persisted but in greatly reduced degree. As an example, Schusterman (1964c) found that the kind of previous training given had little effect on performance in the learning-set situation. This was of theoretical importance because it appeared contradictory to both Harlow's error-factor theory and transfer suppression theory.

The Radiation Program

Rogers reported that the radiation program was finally terminated as surviving irradiated chimpanzees and their controls were given 1000 r of gamma irradiation to the head and further behavioral and physiological measures were obtained before sacrifice (Annual Report 1963–64).

Physical Anthropology

The primary activity in this area again related to blood groups. For example, Wiener, Moor-Jankowski, and Rogers (1965) studied patterns of inheritance for several blood types and blood factors in fifteen chimpanzee families.

Conclusions Regarding Research

Despite the turmoil present at the laboratory during this period because of the impending move and the breakup of scientific staff collaborations, the scientists present during this last phase of the Orange Park facility's life remained active and published material concerning a variety of topics. Most of the work had its roots in earlier studies at the YLPB. Some of these research themes would be carried forth as substantial research programs in the new surroundings of the Atlanta center. However, the effort devoted to the more biomedical approaches that were occasionally present in Orange Park would soon surpass the psychobiological perspectives that dominated work at the YLPB.

THE MOVE

The crowning event of the Bourne Years was the move of 340 miles to Atlanta. The major problem, of course, was how to move the animals. The initial plan was to move the animals by air. This plan was later scrapped partially because of unwillingness of the Navy to provide the transportation. Therefore, the animals were moved by road. An air-conditioned moving van with two drivers was rented and animals were moved in July of 1965 (e.g., Annual Report 1965–66; Bourne 1971). The van was loaded in the early evening and driven throughout the night in order to arrive in Atlanta at around 4:00 A.M. That way the move could be made during the coolest hours of the day. The van was then driven back to Orange Park during the day in order to receive another load of animals the next night. Staff members equipped, with tranquilizer guns and a pistol, in the event of an escape or road accident, accompanied each load of animals in trucks. As many as fifteen-twenty chimpanzees and several monkeys were taken at a time in each van. The whole move was completed without incident and with no noticeable illness or injury to the animals.

Some of the details were recalled in a series of reminiscences published in *Inside Yerkes* (Orange Park 1991). Johnnie Callahan, a vehicle operator at the YRPRC, recalled that at one rest stop the animals made so much noise that people began to gather at the truck and they shortened their stay.

In an audio tape, William A. Draper (1994) recalled one incident:

> about half way there, it was in the evening, it was about half way there, the van pulled off to a truck stop . . . to get something to eat. It was fairly late at night. And we went in and I went in with the truck drivers and I turned around and I

couldn't find the handlers. I said where'd they go. And the truck driver says don't worry about them, they go around the back and they eat in the kitchen. So I said, well that's alright, that's the system, fine, that's okay. So we had our meal in the front and had a reasonably good meal; it was standard fare. Then it came time to leave and one of the truck drivers said you better go back and get those guys, we're ready to go. So I went back into the kitchen and there was a table there and it was a feast for a king, if I ever saw anything amazing. Four or five Black handlers who were just being served by the Black cook absolutely the best food that I had seen all the time that I was in the South. So I said to them, if I had known in any way that you folks were eating like this out here I would have been back out here in a flash to eat with you. And we laughed over that.

Betty Green (1994), a lifelong resident of Orange Park who had worked briefly at the laboratories and had been one of those infected with hepatitis reflected: "I remember thinking that we were losing a prestigious part of our community at the time because of all the doctors that came there and the research that was going on. I knew, even though I was no longer connected with it, and had only a minor connection, I realized Orange Park was losing a valuable piece of itself."

16
Wrapping It Up

In this concluding chapter I discuss briefly the events in both Orange Park and Atlanta after the move of the YRPRC and try to provide some perspective on the history of the YLPB and its significance.

After the Move

In Orange Park

As noted, once it was clear that the Laboratories would be moved, efforts to use the Yerkes Laboratories buildings for a new primate facility continued, but to no avail. With the move, the buildings that Robert Yerkes and his successors had worked so hard to develop were abandoned. A 1976 article in the *Florida Times-Union* noted "newcomers, unless they hit upon an amateur historian, may puzzle over these large beige buildings, now vandalized, and wonder what a small colony of chimpanzees and scientists once did there" (Rhea 1976, B7). The article began on a more optimistic note, however: "in recent weeks the overgrowth shrouding the Yerkes Laboratories of Primate Biology has been stripped away, giving Orange Park a glimpse back to the day it sheltered a significant scientific community."

The grounds remained vacated for about a year after the 1965 move. "For Sale" signs were in place but the designated caretaker was unable to maintain the buildings and they were vandalized. Developer Marvin Wilhite, who had come to Orange Park in the Navy in 1954, purchased the property from Emory (Wilhite 1994). After getting possession, Wilhite tore down some of the buildings and renovated others. By 1976, Wilhite had developed the Foxwood subdivision, a housing development, on most of the land that had not included the actual laboratory buildings. In 1976 Wilhite decided to convert the ten-acre-fence-enclosed area, that had been the functional part of the YLPB to multipurpose use. He first renovated the original Administration building and moved his office inside. A new building was constructed for use by the Orange Park Chamber of Commerce.

Today, one can visit the site, now the Foxwood Center, and see some of the original buildings alongside others that have been added. As of the summer of 2002, seven of the original buildings were still standing. The old caretaker's house was now the Granary, a health-food store (1738 Kingsley Avenue). The old administration building (1712 and 1714 Kingsley Avenue) housed, among other things, a rail transport office and a transcription service. Tenants of the old laboratory building (1728) included a pipe company, a family center, an insurance firm, an accounting office, and a medical supply firm. Marvin Wilhite's real estate firm, AHPLA, occupied the old service building (1718). The original caging facility, or quarters building, (Q Building; 1722) was used for storage. The old gate house (1740) housed offices of American Express Financial Advisors. The old three-car garage (1742; now termed "Serenity: The Cottage") housed a spa whose brochure offered such services as facials, glycolic acid treatments, and bikini hair removal. One can only wonder how Robert and Ada Yerkes would have reacted!

The new buildings house, in addition to the Clay County Chamber of Commerce, apartments, a book seller, a construction firm, and an office of the American Red Cross. The tiny house, located further down Kingsley Avenue, in which the Hayeses had lived and reared Viki, was demolished recently.

Legends about the Laboratories persisted even after its redevelopment. According to a 1992 Halloween article (Avery 1992):

> If you drive by the old Monkey Farm on Kingsley Avenue this Halloween night, you might just hear the ghosts of those chimps screeching or catch a hazy glimpse of the crazed scientist hard at work in his laboratory. But only if you look closely and use a little imagination.
>
> If you're looking for a place to haunt this fright night, stop by the old Yerkes lab and take a look for yourself. but beware of creatures out there monkeying around. (Avery 1992)

Old legends die slowly.

Thomas Wolfe wrote that you can't go home again. William Draper was one of the scientists who did return to Orange Park during a sabbatical in the 1980s. In an audiotaped statement he recalled:

The Administration Building in Orange Park as it looked during the early years of the twenty-first century. Photograph by the author.

> The main buildings were there, a lot of the cages were gone from outside . . . the gate was open . . . and it was a reasonably well kept affair, reasonably well kept building. And inside there was a consulting firm . . . And I also met a number of the secretaries and business manager, whatever. And I explained to them who I was and why I came back, and they said oh my goodness, it's really good to see you because we had heard about this place where we have our offices in, but we've never seen anybody who actually worked here. So I went back with them, back through the main part of the building where I'd first gone in 1961 with Arthur Riopelle . . . not a pleasant experience for me. So I left. (Draper 1994)

In Atlanta

The move to Atlanta was completed in July 1965. The Board of Scientific Advisors met there on October 26 and the building dedication occurred the next day. At the dedication, Adolf Schultz, a leading anthropologist and primatologist, gave the keynote address. Other talks were given by Emory's president Sanford S. Atwood, Ernest M. Allen, Long of the Board of Scientific Advisors, and Director Bourne. Attending the ceremonies were Robert Yerkes' two children, Roberta and David. It was not until 1966 that the main building was finally handed over from the contractor to Emory. The final cost was reported as $1,873,963.

February 1966 saw a site visit to evaluate the new scientific and financial program. The committee filed an unfavorable report; problems cited included funding difficulties, problems of administrative control, and a failure of scientists other than the psychobiologists to study the unique colony of great apes. Bourne maintained that the difficulties were due mainly to inadequate funding. A second site visit occurred in May and resulted in a recommendation in favor of the $603,000 base grant plus a supplement.

The changed emphasis with a reduced percentage of effort directed at psychobiological work and an increase in biomedical investigations was evident early on. Among the highlights of the young program in Atlanta were studies of atherosclerosis, muscular dystrophy, the histology of neurosecretion, the distribution of enzymes in primate brains, hematology, and immunology. Much of the subsequent early history of the YRPRC in Atlanta has been summarized by Dukelow (1995).

The staff members had mixed feelings about the move to Atlanta. The field station provided Bernstein with greater opportunity for the study of animals in groups than had been afforded in Orange Park. Rogers (1994) was concerned that the new facility was all concrete and never fully dry. In Orange Park, the animals could look out at other animals and people moving around all day; their view in Atlanta was impoverished.

Draper (1994) noted the changed atmosphere. He recalled "the change from an earlier mode of doing research, and a simpler, more relaxed yet highly efficient, and effective and productive way, to a more modern, mainstream, jet set kind of approach to things."

Jimmy Roberts noted that the staff members were less involved in the day-to-day aspects of research in Atlanta. In Orange Park there were fewer technicians and the scientific staff made most of their own observations. In Atlanta, there was more help that was relied on for some of the basic collection of data.

Because this book is a history of the Yerkes Laboratories in Orange Park, detail concerning its subsequent history in Atlanta would be inappropriate. However, a small sampling of what happened after the move is relevant.

When Bourne retired as director in 1978, he was succeeded by Frederick A. King, then of the department of neuroscience of the University of Florida. He specialized in brain function. In September 1994, King, in turn, was succeeded by Thomas R. Insel, M.D. Among his interests were research on obsessive-compulsive disorders in humans and pair bonding in voles. He resigned in 1999, to be succeeded by an interim director; Insel became the director of the National Institute of Mental Health in 2002. In September 2001 Stuart Zola, a physiological psychologist noted for his research on learning and memory, was named the director.

The annual Progress Report for 1989 ran fully 308 pages. There were twenty-one doctoral, fifty technical personnel, twenty-two people in administrative and support services, thirty-seven animal care employees, and 136 adjunct faculty members. I counted a total of 130 separate projects. The report listed over $7,300,000 in individual grants, of which over $6,390,000 came from the Public Health Service.

By 2001, there were four major research disciplines (microbiology and immunology, neuroscience, psychobiology, visual science). Four separate, but closely interlinked, research centers were also housed at the Primate Center (the Emory Vaccine Research Center, The Center for AIDS Research, Science and Technology Center in Behavioral Neuroscience, and the Living Links Center, established in 1997 by Frans de Waal). According to a press release, the annual budget for the YRPRC now stood at $36 million. This budget can be put in perspective by comparing it to the last year in Orange Park for which data are available—1962 with a total budget of $395,613. In absolute dollars, this represents an increase by a factor of 91. Considering inflation, this still represented an increase by a factor of 15.5.

As of April, 2002, the base breeding colony included 82 mangabeys, 1,024 rhesus monkeys, 247 pig-tailed macaques, and 3,690 rodents. The research colony included 30 capuchin monkeys, 198 mangabeys, 1 stump-tailed macaque, 18 crab-eating macaques, 1,256 rhesus monkeys, 144 pig-tailed macaques, 154 chimpanzees, 40 baboons, and 54 squirrel monkeys. Thus, the shift was toward more monkeys from species that are useful in biomedical research—the direction suggested by several of the evaluators of the YLPB during the 1930s.

In June and July of 1968 the YRPRC hosted the Second International Congress of Primatology. It was presented as "A Memorial Congress in Honor of Robert M. Yerkes." Leonard Carmichael gave a presidential address with an emphasis on the history of the Yerkes Laboratories and the career of Robert Yerkes (Carmichael 1968).

In October 1976, the YRPRC sponsored a conference celebrating the centennial of Robert Yerkes' birth. The first of four programs dealt with "Historical Beginnings of Research on Great Apes" and included numerous YLPB alumni. It was chaired by Austin Riesen and included: Roberta Yerkes Blanshard and David Yerkes, and such Yerkes associates as Meredith Crawford, George Haslerud, Myrtle McGraw, James Elder, Shirley Spragg, Harold Coolidge, Vincent Nowlis, and Karl Pribram.

The YRPRC was a prime target for the movement opposing animal research that matured during the 1970s and 1980s. Organizations such as "Last Chance for Animals" charged that the Center wasted funds, cured no diseases, and functioned only to line the scientists' own pockets. Break-ins were threatened. In response, security at the Center was greatly increased and some staff members went on the offensive. King became a leading advocate for animal research. The activists attacked Zola even before he arrived in Atlanta.

A group known as the Committee on Animal Models in Biomedical Research published a critical analysis of the YRPRC (Committee on Animal Models 1995). The report challenged the view that research on nonhuman primates has beneficial effects for human health.

Some Recent Follow-Up Research

Many of the research projects from the YLPB in Orange Park have been followed up, extended, and superceded since. That is the way of scientific investigation. The field of primatology has grown enormously and this is not the place for an extensive review of subsequent developments. Nevertheless one can point to a few areas where the YLPB research can be seen in contemporary context.

Work following up Yerkes' and others' early studies of sexual behavior has been extensive. Working at the YRPRC, Ronald Nadler conducted an extensive program on the sexual behavior in apes, including broad comparative studies of sexual behavior in chimpanzees, gorillas, and orangutans (e.g., Nadler 1977). Nadler (1981) summarized research conducted during the whole span of Yerkes Laboratory life on the ontogeny of sexual behavior, estrus, and the hormonal regulation of sexual behavior in apes. In addition, Nadler, Dahl, Collins, and Gould (1994) followed up Yerkes' earlier work on male versus female regulation of sexual behavior in chimpanzees in relation to female endocrine status, dominance relations, and social compatibility.

The Temerlins (Temerlin 1975) expanded upon the work of Kellogg and Kellogg, Hayes and Hayes, and Finch by raising a chimpanzee, Lucy, in their home. The unique aspect of this study is that the Temerlins were able to raise Lucy together with their son Stephen for a period of 10 years.

There has been recent interest in cooperative problem solving as studied at the YLPB by Meredith Crawford and his associates. Mendres and de Waal (2000) studied cooperation with variations on Crawford and associates' earlier methods. There also have been many studies of imitation, effects of early experience, social behavior and organization, and many of the other topics studied in Orange Park.

Perhaps the most visible development of recent decades has been the reemergence of interest in animal cognition, especially that of nonhuman primates (e.g., Griffin 1984; Tomasello and Call 1997). Robert Yerkes was one of several psychologists who retained their interest in cognition at a time when it was out of fashion in many circles (Dewsbury 2000). He would be pleased by recent developments.

The pioneering research done at the YLPB helped to stimulate interest in comparative studies of apes. As is almost always the case in science, not all of the results have held up well. The stimulative effects of this work have been substantial, however.

Perspectives

It is appropriate to try to provide some overview and closure for the story of the Yerkes Laboratories of Primate Biology as they existed in Orange Park, Florida during 1930–65. This revisits some of the themes introduced in Chapter 1. One must wonder whether when as a graduate student Robert Yerkes envisaged a remote facility, he could have realized the tortuous path it would take and the number of lives that would be affected by it.

Individuals

THE DIRECTORS

An interesting theme in the history of science concerns the relative roles of the individual scientist and the surrounding milieu in effecting scientific change. As shown here, and elsewhere, both are important.

One can start with the five directors. Robert Yerkes deserves full credit for his vision in conceiving of this facility and for the persistence displayed in bringing it to reality. Yerkes was a planner who worked tirelessly to bring his plans to fruition. He must be credited with showing how well chimpanzees can be maintained and bred in captivity and with overseeing substantial research. A shy man, he moved well enough among the movers and shakers of his time to secure the necessary resources to establish the laboratories. He was less successful as an administrator. His style was authoritarian and he seemed to drive away those who departed from his ways of thinking. Further, as the prevailing conditions at the Rockefeller Foundation changed, he found himself out of touch and out of favor with the survival of the YLPB at risk. Finally, he was out of the position as director.

Karl Lashley had an air of confidence that Yerkes lacked. Lashley presented himself, and was presented by his colleagues, as the archetype of the disinterested scientist, living for the pursuit of truth. Weidman (1999) believes this to be illusion, claiming instead that Lashley's science was driven by political motives. In my view, these claims are unsupported (Dewsbury 2002a). Lashley was driven by a thirst to understand the brain and its relation to mind. He lacked Yerkes' conviction of the scientific value of chimpanzees as laboratory animals. Thus, he used the YLPB as a convenient home within which he could conduct research. He seemed determined to infuse the YLPB with his values but left the day-to-day running of the laboratories to Henry Nissen. Lashley was not threatened when surrounded by creative scientists; he sought them out and allowed them the latitude that he wanted for himself. This made for a very different environment from that under Yerkes. Surely, it was Lashley of the five directors who had the highest standing as a scientist among his peers.

Nissen was more gentle. He was the archetypical number-two man, always in Yerkes' or Lashley's shadow. It is a mistake to underestimate the qualities of the man, however. He probably knew more about chimpanzees than any other human of his time. His science was so solid as to earn him election to the National Academy of Sciences. He was a humane man who genuinely loved working with the chimpanzees and had a unique rapport with them. Nissen was not cut out to be a director, however. The job, together with events in his personal life, were too much for him.

Arthur Riopelle was a combination scientist-administrator. Times were changing. With federal grants of increasing size available for active and visible scientists, there was a new emphasis on both publication and funding. Riopelle was able to bring some productive scientists to Orange Park and to work effectively with them. A hands-on administrator, he was also involved in research activities. Where Riopelle had difficulties was in his relations with the Emory administration. Whereas the Yale administrators had solidly supported Yerkes, at least in the beginning, the Emory administrators, working in a different context, saw it as an appendage to be run for the benefit of their university. Riopelle was viewed as a bit of a maverick in placing the interests of the Orange Park facility above those of Emory. When the Regional Primate Center program was established, he came into direct conflict with the Emory administrators, lost and left. It is doubtful that any administrator could have kept the YLPB in Orange Park, but Riopelle's strong stands made the process especially difficult.

Geoffrey Bourne, despite his extensive publication record, may have been the purest administrator of the group. He came from Emory and operated the YRPRC with the goal of integrating it into the Emory family. He was a showman and a salesman who was at ease with the movers and shakers in a manner that was foreign to Yerkes, of little interest to Lashley, and intimidating to Nissen. Bourne brought the laboratories to Atlanta and facilitated the shift to a more biomedical orientation.

THE SCIENTIFIC STAFF

As important, in some respects, as the directors were those who actually conducted the research—the scientific staff. The story of the people of the YLPB sometimes reads like a soap opera. There were the suicides of Henry Nissen and Richard Davenport, the drowning of Robert Blum, and the death while skiing of Paul Schiller. There were births, marriages, and divorces, as in any human community. Recall that Falk (1994) spoke of the "Tennessee Williams dynamic" in Orange Park and Warren Weaver (1955) termed the Hayes-Nissen saga "Desire Under the Palms." There were important conflicts, such as those of Yerkes with both Bingham and Tinklepaugh, Nissen with Ferster, and Bourne with Jankowski. Surely, individual egos in addition to legitimate scientific issues, affected the progress of the YLPB just as they affect all such institutions. It is hard to know whether this sample differed from humanity at large.

Cathy Nissen (1955) exclaimed "What a variety of people have passed thru these halls!" Scientifically, there were those promoting various versions of a hard-science approach, such as Spence, Ferster, and Kelleher, and those more comfortable with a softer approach, such as Nissen, Rogers, and Yerkes himself. Yerkes and Nissen sought an integrative psychobiology; Lashley's was a reductionistic approach. Others adopted a wide range of diverse strategies.

Of course, the majority were typical, hard-working scientists. They are not the subjects of colorful tales. Other personalities ranged from the extroverted Davenport to the self-effacing Nissen. The staff included one Nobel Prize winner, Sperry, and 4 APA presidents, Yerkes, Lashley, Hebb, and Hilgard (in addition to Leonard Carmichael).

THE SUPPORT STAFF

The invisible people were those on the support staff. The lowest rungs were occupied almost exclusively by African Americans, as was typical of the south at that time. Much of the office help was female. Murray Townsend was an especially important player, filling in with needed administrative work when the attention of directors was elsewhere. One key to the success of the laboratories was the ability to retain long-time superintendents—William Atwater and Graham Townsend.

THE ANIMALS

The animals served through coercion. It would be difficult to look into the eyes of a chimpanzee and not see a sensitive, thinking mind. Yet, the hard-nosed behaviorist would maintain that no such mentalistic constructs are necessary to explain chimpanzee behavior. Can one hear worker after worker tell of the propensity to plaster humans with feces and spray them with water, about the only materials at their disposal, and not see some sense of either humor or retribution? Are these simply the mechanical acts of a machine acting according to the strictest of classical behavioristic principles?

Animal welfare issues were much less prominent during this period than today and the housing conditions were substandard by contemporary standards. However, there is every indication that the vast majority of the researchers genuinely cared about the animals and their well being.

PERSONALITY AND SCIENCE

The conduct of science is affected not only by the surrounding milieu, but also by the personalities and proclivities of the scientists. Robert Yerkes had a strong sense of family and tended to view chimpanzee families as constructed in a manner similar to the model human families that he knew and admired in his place and time. He cannot be blamed for being uninformed regarding field work done since his death showing that this is not the case. However, his approach does provide an instance of the impact of personal values on scientific positions. Surely, his views on dominance and the granting of privilege between males and females appears to have been shaped by his own views on issues of gender in humans. The very idea for the YLPB emerged from his commitment to service and belief that through science humans could rationally control both society and nature to improve the human condition.

I have speculated elsewhere (Dewsbury 2002a) that Lashley's science may also have been a reflection of his personality. His misanthropic tendencies may have contributed to his emphasis on basic, rather than applied, science. His egotism may have affected his attitudes toward the theories of others. He was a contrarian in many respects in and outside of science. Lashley was extremely independent, determined to go his own way regardless of those around him. His science emphasized internal, control with a lesser role for external influences than in the models of some of his contemporaries. The possibility that his science reflected his personality merits further study.

CHANGE

Lives are stories of transformation. As we grow we change and so does our surrounding environment. The 35 years of the existence of the Orange Park facility were years of great change in society and in science. There was cultural change in individual relationships, race relations, and governmental structure. There were scientific changes in paradigms, problems, and funding. In its life the institution must change. In a sense, its success is a function of how it adjusts to change. Yerkes was a man of consistency who did not handle change well. Lashley insulated himself from change with the tenacity of his vision of science. Nissen followed, as he often did, in Yerkes' footsteps. Just as Riopelle was beginning to bring fundamental change to Orange Park, the facility was taken from him. A man of another culture, Geoffrey Bourne shaped a new institution that incorporated only some of what had been in the old.

The Scientific Milieu

CHANGING TRENDS

Research programs occur within a scientific milieu. Yerkes was a product of an earlier time when William James, Josiah Royce, and Hugo Münsterberg ruled psychology at Harvard. His was a psychology of conscious experience and of mental processes. If he was out of tune with Harvard's philosophy department and president while a faculty member there, he was well-tuned to developments in comparative psychology at the national level. The YLPB was founded during a period when Yerkes was greatly respected. As the more reductionistic sciences came into prominence during the 1930s, he came to be viewed, at least in some quarters, as an old-fashioned scientist. Lashley was a man of that time. Riopelle and Bourne also fit with their times. They were scientist-administrators at a time when cooperative research was increasingly valued.

The influences of trends at a less global level are also apparent. The influx of operant behavioristic psychology occurred during the 1950s as the influence of B. F. Skinner was growing. The use of arousal theory in interpreting results was prominent during the 1960s in psychology broadly and in Orange Park. Further, this was a time of interest in motivational systems other than food, water, and sex; this was reflected in some of the YLPB projects. During that period there also was a resurgence of interest in field studies reflecting the impact of such scientists as Louis Leakey and Sherwood Washburn and this was reflected in the work at Orange Park as well.

There were many changing patterns in the conduct of science. Yerkes was a scientist who relied on his insight and displayed a remarkable willingness to cast aside the results of statistical analyses that ran counter to that insight. Such procedures would be unthinkable in the scientific context of the Riopelle years. Scientific articles became sprinkled with the results of *t*-tests and analyses of variance and scientists ignored their results at their peril.

There is a long-standing split in animal behavior studies between devotees of laboratory versus field research. Most scientists see the advantages of each. Where careful control of conditions is essential to a project, laboratory research is required. Where, however, ecological validity is important, the field research is essential. Yerkes understood the value of field work and enabled Carpenter, Nissen, and Bingham to conduct field studies. However, in the prevailing milieu he was unable to sustain his program and the generality of some of the laboratory results was limited because of it. It was only when conditions changed during the 1960s that YLPB scientists would again venture into the field. Carpenter, among others, was important in this shift.

PATRONAGE

As noted throughout this book, the effects of changing patterns of patronage are clearly illustrated in the history of the YLPB. In Robert Yerkes' day, one made social contacts and worked them in efforts to secure funding. Patronage was from private foundations and one approached them to advance one's project. The period from 1930 to 1965, the life of the YLPB, coincides almost perfectly with the dramatic shift in patronage from private to federal sources. The new scientist-administrator had to write grant proposals to, and host site visits from, government officials and their representatives. Private sources contributed 95% of the funding during Yerkes' years as director; federal sources provided 86% of the funding for the Riopelle years—a dramatic turnaround in a relatively short period of time (Table 9.1).

The changing funding patterns had effects not only on the fund-raising activities of the directors but on the way in which they could run the scientific programs. The base support from such agencies as the NSF and NIH was important but it was necessary to secure individual grants for specific projects in addition if budgets were to be met. This greatly limited the flexibility of the directors to allow good people to find their own directions and to shift interests with changing short-term changes in conditions.

What greater example of the impact of these changes in patronage could one find than the radiation project? This was a program supported by government funding in the context of the cold war nuclear concerns. It revealed few, if any, direct effects of radiation on behavior. Given contemporary values, it is easy to be critical of this program on ethical grounds but it should be understood in its post-World War II context and the fear of nuclear war and effects of nuclear power.

PROFESSIONALIZATION

Many of the changes that occurred during this period reflect the professionalization of science. The iconoclastic scientist was being replaced by the scientist-administrator who was effective in working with groups, managing budgets, and promoting research in addition to doing good science.

Yet another aspect of successful science that is revealed in the work in Orange Park is the role of promotion of one's research. Such staff members as Nissen, Rogers, and Davenport, were minimally concerned with the selling of their science and had correspondingly less impact on the field. The master salesman was Harry Harlow, who was an engaging speaker and a writer with a flair for turning a phrase that would bring notoriety, and sometimes criticism, to his work. In areas such as "contact comfort" and "learning to learn" the YLPB scientists made appreciable early contributions. Some may have been in advance of Harlow's. But it is Harlow's name that is associated with the work. The research must not only be sound but must be communicated and marketed.

THE HEGEMONY OF MEDICAL RESEARCH

Federal support for science is dependent, in part, on payoff. The Congress seeks evidence of societal benefits. In the life sciences, this usually means benefits to human health. Although the NSF was created to facilitate the development of basic science, even there administrators would fight ongoing battles with those convinced that applied science could develop in the absence of a foundation of basic science. In the area of the life sciences, it was generally relatively easy to garner support for research that would cure a disease. It was much harder to get support for research that would increase basic scientific knowledge. It was this influence that had a major effect on the demise of the Orange Park facility.

THE ETHICS OF ANIMAL RESEARCH

If asked today to list significant issues related to research with nonhuman primates, those concerning the ethical treatment of animals would be high on the list. However, issues of ethics are virtually absent throughout the history of the YLPB in Orange Park. One finds virtually no protests regarding captivity and handling and there were no pickets outside of the facility's famed barbed-wire fences. That is dramatically different from the history of the YRPRC in Atlanta, which has served as a lightning rod for protests about animal research.

There were some high-profile protests against the animal research of psychologists, such as John B. Watson, Ivan P. Pavlov, and Edward L. Thorndike, near the beginning of the twentieth century (Dewsbury 1990). However, this was followed by a period of relative calm, with only an occasional minor skirmish around such issues. In 1925 the APA formed its Committee on Precautions in Animal Experimentation with Robert Yerkes as the founding chair of the committee (Dewsbury 1990).

Many scientists, including those who have worked at the YLPB and YRPRC, are more conflicted about these issues than might be assumed by the general public. Mason, for example, presented a perspective that would be shared by many scientists: "I have never been able to reconcile fully the conflict between my interest in understanding the world as it is—that is to say, in science—and my feelings and moral attitudes toward the animals I work with. On the one hand, I believe implicitly in the value of the scientific approach. On the other hand, I believe just as firmly in the inherent value of the animals I work with and their right to a free and independent existence" (Blum 1994, 101–2). Today, we cannot read about the activities of the YLPB without thinking about these issues. What must be understood, however, is how little debate there was about them during the Orange Park years.

WRITING SCIENCE

The work of Yerkes and his associates on male-female dominance and privilege provides a classic example of the role of primary, secondary, and tertiary simplification in the reporting of scientific results (Dewsbury 1998b). Qualifications and limitations that are specified in primary data reports may be deleted as data are summarized and reported in more abbreviated forms. The results often take on an appearance of reality that is unwarranted by the supporting data. Although Yerkes' work in this area provides a prime example, it is by no means unique and illustrates the broader phenomenon of simplification.

The Broader Milieu

ORANGE PARK

Descriptions of life in Orange Park are remarkably consistent throughout the YLPB's history. The heat, insects, and water were sources of irritation. More important, many of the highly trained scientists from northern urban environments experienced some culture shock when being placed in a small southern town as it existed in the early years. Above all, it was the isolation from other colleagues and some of the amenities of the city and university that most affected the scientific staff members. Living conditions outside of the laboratory affect scientific work just as they affect work in other walks of life.

URBANIZATION

Ideally, a primate station is located in a remote area so that the animals don't bother the people and the people don't bother the animals. When Robert Yerkes selected a site for the YLPB, it was down an unpaved road in a small southern town with only a few commercial businesses. As the town of Orange Park and the surrounding area grew, more and more attention was paid to the facility. Although there were reservations about moving the YLPB to the urban location in Atlanta, it could be meaningfully argued that Orange Park was becoming just as urbanized. It is interesting that the same process has occurred with the Lawrenceville field station of the YRPRC that was founded to enable the study of social groups of primates away from urban Atlanta (Blum 1994).

RACE AND GENDER

The period of the YLPB in Florida corresponds with a dramatic change in race relations in the United States in general and the South in particular. Throughout its history, the YLPB was stratified by race. The scientific staff and most, if not all, of the office workers were white; the caretakers and laborers were generally black. During the early years of the laboratories, one finds relatively few comments on this phenomenon. The first two directors had views on race that can be regarded as racist by contemporary standards (Dewsbury 1993; Yerkes 1923a). It is hard to know how widespread in the YLPB these attitudes were.

The 1950s and 1960s saw the beginning of change in race relations and dramatic reductions in at least some aspects of segregation in the South. It was in the interviews from this time that some scientists expressed discomfort with the status quo. Scientists such as Berkson, Bernstein, Draper, and Schusterman all addressed this issue in one way or another. They expressed discomfort with the racial situation and appear to have tried to subvert it in small ways in their personal relations. Perhaps, with raised consciousness, they expressed what others felt.

I found no evidence of racial strife or protests at the YLPB. The main effects may have been in the lives of the scientists outside of the laboratories and in the loss of potential talents that might have been developed had gifted African Americans been given better positions.

Issues of gender were present as well, although perhaps no more than elsewhere in the science of the time—at least after the departure of Robert Yerkes. Yerkes made it clear that there would be no women on his scientific staff. In later years numerous women worked at the laboratories. Surely, the support staff was sharply divided by gender.

BROAD PHILOSOPHIES

Of course, work at Orange Park was influenced by the broader cultural context as well. One example lay in the progressivism that characterized both Robert Yerkes, with his plans for social engineering, and the Rockefeller Foundation. This confluence of values facilitated the very founding of the YLPB.

Robert Yerkes was committed to the ideals of practical idealism, progressivism, and culture with which he grew up. This world view was challenged in a period of rebellion around World War I. The simple formulae, like the agrarian past, were rejected in an era marked by the depression of 1914 and the horrors of world war. The psychologies of Sigmund Freud, Havelock Ellis, and John B. Watson helped lead many intellectuals away from the traditional culture in which Robert Yerkes was mired. This period has been characterized as an end of American innocence.

At this time, the received mentality was under siege from young intellectuals: "Sexual and family relationships . . . were contrived and culturally determined, not ordained by God or by an unchangeable human nature. They could be altered and molded to enhance human happiness . . . They were survivals from an unenlightened age and generations of social dysfunctions, mental illness, and crabbed, unhappy, and maladjusted human beings" (Hawley 1992, 113). Robert Yerkes was a product of the earlier time, out of sync with this new perspective. Indeed, he was so deeply committed to the ideal of the human family that he imposed it upon his conception of social structure in chimpanzees and would see in them many of his traditional values.

WORLD WAR II

The impact of World War II was large because of both the loss of personnel and resources and the inflow of military personnel to the Orange Park area. A facility that was designed to train chimpanzees housed a program to train human radar operators. The period immediately following the war and into the cold war was one of tensions and concerns about nuclear energy and the threat of nuclear war. Radiation research became an important program in Orange Park.

Overall Evaluation

At the bottom line, we must ask questions about the overall impact of the thirty-five years of the YLPB in Orange Park. After all, the facility was established for the conduct of research and it is the products of that research that must remain foremost. It is natural to want to evaluate it in terms of contemporary standards although the only fair evaluation would be in relation to the standards of the day.

For those who see justification for animal research only in terms of benefits to human health, the direct gains must seem quite modest. There was no major disease cured as a result of the research that was conducted there. However, in contrast to the YRPRC in Atlanta, there was very little work of a biomedical nature that was conducted. Surely, Robert Yerkes' plan of utilizing chimpanzees to gain knowledge for a program in social engineering did not come to pass. The emphasis during most of the thirty-five-year-history of the YLPB was on basic research—that was designed to advance knowledge rather than to solve an immediate problem. In that respect, the activities of the YLPB fare a good bit better. The facility provided an important foundation for the vast knowledge of nonhuman primates and their behavior that we have today.

One must go back to the time of Yerkes' founding of the laboratories, as in the book of Yerkes and Yerkes (1929), to see how little was known about the great apes. Today, hardly a day goes by when chimpanzees cannot be seen on some television channel and some primate researchers have become media personalities. Obviously, the facility was not solely responsible for this remarkable surge of interest in primate research. However, it is fair to suggest that it was an important cog in setting the conditions for this explosion of interest. Haraway (1989) called Yerkes' founding of his "paradigmatic laboratory" his "enduring contribution to science" (61).

Contemporary historians, influenced by postmodern and social constructionist perspectives, question such concepts as progress in science. Even the notion of the "fact" is subject to debate. Granting that facts do not exist apart from observers and can mean different things in different discourse communities, the more than seventy years since the founding of the YLPB have seen an enormous increase in what we know about primates and their behavior. This encyclopedia of information has changed how we think about primates—both human and nonhuman—and how we think about such issues as gender, dominance, and power.

Each of the five administrations had its highlights. Clearly, the major contribution during the first administration of the YLPB was in showing that chimpanzees could be maintained, bred, and studied on a comparatively large scale in a captive environment. It is difficult to overemphasize this accomplishment because it set the stage for much that came later.

Surely, the most controversial research of the time, though it was done in New Haven, was Jacobsen's on the function of the frontal lobes. This work had major implications for human health and led, in the hands of others, to many lobotomies performed on human patients. Unfortunately, the results were not beneficial overall. On a more positive note, Spragg's work on morphine addiction became a classic in psychopharmacology and was one of the founding experiments in what has become a very large field.

It was during the Yerkes years that a whole series of investigations provided basic information about the instinctive behavior, social behavior, and sensory and motor processes of chimpanzees. Crawford's research on cooperation in problem solving was among the highlights.

It was in the early years of the YLPB that important pioneering steps were taken toward establishing field research on primate behavior as a viable endeavor. The long-term program of research on behavioral development, that would continue into the years in Atlanta, was begun during this period as well.

The work of Yerkes and his associates on the sexual cycles and behavior of chimpanzees opened up a field that has been extremely productive. One can be critical of some of their methodology and conclusions, but it is hard to deny the role in establishing sexual function as a legitimate field of study.

Yet another important aspect of research during this period concerned animal cognition. Yerkes and a few others kept alive and advanced the study of animal cognition during the 1930s. This was a contrarian approach relative to much that was happening in the field but its links to current perspectives should be clear.

Finally, during this period, as in later periods, the YLPB was an important training ground for scientists who would go on to have major impact elsewhere. Part of its success lay in the education and training of scientists such as Meredith Crawford, Carlyle Jacobsen, Winthrop Kellogg, Henry Nissen, and Kenneth Spence.

It was during the Lashley years that the long-term infant studies program became a major endeavor. This program contributed in an important way to the understanding of the complex processes that underlie development in our closest relatives. The work included not only the study of normal development but the effects of unusual environments on those processes. It is this program, perhaps more than any other, that characterizes the work in Orange Park.

The work of Karl Lashley and his associates on brain function in primates that was conducted at the YLPB is among the stellar accomplishments. In addition, some of the skills and techniques that Roger Sperry developed in Orange Park would play a role in his earning a Nobel Prize.

Yet another of the influential psychologists of the century was Donald Hebb. In 1949, just two years after leaving Lashley and Orange Park, his *Organization of Behavior* was published. It is there that one finds the so-called Hebb synapse that has become a mainstay of contemporary cognitive neuroscience. Lashley's influence is clear; indeed, he was offered and declined co-authorship. Among other important work were Young's program on the physiology of sexual function and Riesen's on the effect of sensory restriction on later development.

The third period was too brief to generate a list of accomplishments comparable to those of other periods. A major program during this period was that in operant psychology. Although it had little lasting impact in Orange Park itself, use of operant conditioning techniques in studies with nonhuman primates has had a long and effective history since that time that is worthy of a study in and of itself. The infant studies program, the pride of Henry Nissen, continued during this time. Of lesser impact was the radiation program, which came to maturity during these years.

During Arthur Riopelle's administration, the range of species available for study was increased. Studies of basic behavioral patterns and social behavior in several species of primates reached their zenith in Orange Park with the work of Mason and Bernstein. Facilities were developed for the study of social groups, thus adding a whole new dimension to what could be learned from animals in small cages. The ever-present infant studies program, then in the hands of Rogers, Davenport, and others, continued and began to generate more impressive presence in the published literature. A program on aging was added. Spearheaded by Berkson, the program on mental retardation and stereotyped behavior evolved. This had both a basic and an applied focus.

It is unfair to compare the accomplishments under Bourne with those of other eras both because it was relatively brief and because it was a time of transition to the Atlanta facility. Nevertheless, some real progress was made. This was a time of still greater visibility in the published literature. It was during this time that the laboratory personnel were able to return to the Yerkes-Carpenter effort of field research. That was critical in evaluating problems of the ecological validity of the studies conducted in the laboratory. Programs of research on basic behavioral patterns, social behavior, sensory/perceptual function, development, stereotypy and mental retardation were continued and expanded. The foundations of the more biomedically oriented programs in Atlanta were laid during this time.

A Final Word

Lashley (1950c) noted that "the great value of the work here is not in spectacular discoveries but in the steady accumulation of data on systematic problems" (8). The progress of science has been characterized as a wall built one brick at a time. The record of the YLPB is one of many such bricks in the wall of understanding of primate behavior and biology. Scientists in Orange Park studied an array of problems. Their contributions were in building a base of fundamental knowledge, no single one of which stands out as a truly remarkable discovery. The resources of the laboratory were not devoted to the solution of a single, practical problem but rather to the study of a diverse range of biobehavioral characteristics of out closest relatives. In that sense, the program was successful.

As Gray (1955) summarized it: "while there are hundreds of laboratories studying rats, dogs, guinea pigs and other conveniently available organisms, in all the world there is only one laboratory colony of the sensitive, near-human, almost embarrassingly intelligent chimpanzees—in other words, only one Orange Park" (77).

References

Note: Frequent reference is made to the Yerkes laboratories' annual reports and to the minutes of the annual meetings of the administrative boards. These were obtained from the files of the Yerkes Regional Primate Research Center, which have since been housed at the Woodruff Library, Emory University, Atlanta, GA, and from the Robert Yerkes Papers in the Yale University Library, New Haven, Connecticut. Please note that personal communications appear in a separate section at the end of the reference list.

Aberle, S. D., and G. W. Corner. 1953. *Twenty-five years of sex research: History of the National Research Council Committee for Research in Problems of Sex 1922–1947.* Philadelphia: Saunders.

Adams, W. R. 1994. *Historic building survey of the Town of Orange Park, Florida.* St. Augustine, FL: Historic Property Associates.

Allee, W. C. 1933. Gorillas in a native habitat. *Ecology* 14:319–20.

Allen, E., and R. M. Yerkes. 1937, March 3. Progress Report to the Committee on Research on Problems of Sex. Grantee Allen and Yerkes File. Archives of the National Academy of Sciences, Washington, DC.

Angell, J. R. 1928, October 4. Letter to Max Mason. Rockefeller Archive Center, North Tarrytown, NY.

———. 1929, March 20. Letter to George E. Vincent. Rockefeller Archive Center, North Tarrytown, NY.

———. 1930, March 20. Telegram to Robert M. Yerkes. Yerkes Papers. Yale University Library.

———. 1933a, March 22. Letter to Robert M. Yerkes. Yerkes Papers. Yale University Library.

———. 1933b, April 6. Letter to Robert M. Yerkes. Yerkes Papers. Yale University Library.

———. 1933c, April 17. Letter to Robert M. Yerkes. Yerkes Papers. Yale University Library.

———. 1933d, May 31. Letter to Robert M. Yerkes. Yerkes Papers. Yale University Library.

——— 1933e, June 8. Telegram to Robert M. Yerkes. Yerkes Papers. Yale University Library.

———. 1933f, June 12. Letter to Robert M. Yerkes. Yerkes Papers. Yale University Library.

Angier, R. P. 1930, March 22. Letter to Robert Yerkes. Yerkes Papers. Yale University Library.

———. 1933, May 8. Letter to Robert M. Yerkes. Yerkes Papers. Yale University Library.

Anonymous. n.d. Data concerning plan for the study of apes and monkeys. Yerkes Papers. Yale University Library.

Appel, T. A. 2000. *Shaping biology: The National Science Foundation and American biological research, 1945–1975.* Baltimore: Johns Hopkins University Press.

Ash, M. 1995. *Gestalt psychology in German culture, 1890–1967: Holism and the quest for objectivity.* Cambridge: Cambridge University Press.

Avery, E. 1992, October 31. Monkey farm still sends chills. *Florida Times-Union* (Westside Community News), 1.

Bayne-Jones, S. 1937, November 20–21. Record of conference on Yale Laboratories of Primate Biology. Group 1.1, Series 200D. Rockefeller Archive Center, North Tarrytown, NY.

———. 1938, March 3. Letter and memorandum to Warren Weaver. Group 1.1, Series 200D, Rockefeller Archive Center, North Tarrytown, NY.

Beach, F. A. 1961. Karl Spencer Lashley 1890–1958. *Biographical Memoirs of the National Academy of Sciences of the United States* 35:163–204.

———. 1987. Donald Olding Hebb (1904–1985). *American Psychologist* 42:186–87.

Beach, F. A., D. O. Hebb, C. T. Morgan, and H. W. Nissen. 1960. *The neuropsychology of Lashley.* New York: McGraw-Hill.

Benjamin, L. T., Jr., and D. Bruce. 1982. From bottle-fed chimp to bottlenose dolphin: A contemporary appraisal of Winthrop Kellogg. *Psychological Record* 32:461–82.

Bennett, C. F. 1989. *Twelve on the River St. Johns.* Jacksonville, FL. University of North Florida press.

Berkson, G. 1962. Food motivation and delayed response in gibbons. *Journal of Comparative and Physiological Psychology* 55:1040–43.

———. 1963. Psychophysiological studies in mental deficiency. In *Handbook of mental deficiency: Psychological theory and research,* ed. N. R. Ellis, 556–73. New York: McGraw-Hill.

———. 1965. Eye fixation responses to one or two stimuli. *Journal of Comparative and Physiological Psychology* 59:422–26.

Berkson, G., and R. K. Davenport, Jr. 1962. Stereotyped movements of mental defectives. I. Initial survey. *American Journal of Mental Deficiency* 66:849–52.

———, and W. A. Mason. 1963. Stereotyped movements of mental defectives. III. Situation effects. *American Journal of Mental Deficiency* 68:409–12.

———. 1964a. Stereotyped behaviors of chimpanzees: Relation to general arousal and alternative activities. *Perceptual and Motor Skills* 19:635–52.

———. 1964b. Stereotyped movements of mental defectives. IV. The effects of toys and the character of the acts. *American Journal of Mental Deficiency* 68:511–24.

Berkson, G., and S. V. Saxon. 1963. Situation and stimulus effects on stereotyped behaviors of chimpanzees. *Journal of Comparative and Physiological Psychology* 56:786–92.

Berkson, G., and R. J. Schusterman. 1964. Reciprocal food sharing of gibbons. *Primates* 5:1–10

Bernstein, I. S. 1961. Response variability and rigidity in the adult chimpanzee. *Journal of Gerontology* 16:381–86.

———. 1962. Response to nesting materials of wild born and captive born chimpanzees. *Animal Behaviour* 10:1–6.

———. 1964. A field study of the activities of howler monkeys. *Animal Behaviour* 12:92–97.

———. 1965. Activity patterns in a cebus monkey group. *Folia Primatologica* 3:21–24.

———. 1966a. An investigation of the organization of pigtail monkey groups through the use of challenges. *Primates* 7:471–80.

———. 1966b. Analysis of a key role in a capuchin (*Cebus albifrons*) group. *Tulane Studies in Zoology* 13:49–54.

———. 1967a. Intertaxa interactions in a Malayan primate community. *Folia Primatologica* 7:198–207.

———. 1967b. A field study of the pigtail monkey (*Macaca nemestrina*). *Primates* 8: 217–28.

———. 1967c. Defining the natural habitat. *Proceedings of the First Congress of the International Primatological Society,* 177–79. Stuttgart: Gustav Fischer.

Bernstein, I. A., and W. A. Mason. 1963a. Activity patterns of rhesus monkeys in a social group. *Animal Behaviour* 11:455–60.

———. 1963b. Group formation by rhesus monkeys. *Animal Behaviour* 11:28–31.

Bernstein, I. A., and L. G. Sharpe. 1965. Social roles in a rhesus monkey group. *Behaviour* 26:91–104.

Bingham, H. C. 1926. Some notes on chimpanzee feeding September 21 to November 15, 1926. Papers of the Yerkes Regional Primate Research Center. Woodruff Library, Emory University, Atlanta, GA.

———. 1929a. Selective transportation by chimpanzees. *Comparative Psychology Monographs* 5 (4): 1–45.

———. 1929b. Chimpanzee translocation by means of boxes. *Comparative Psychology Monographs* 5 (3): 1–91.

———. 1930a, March 1. Letter to Robert Yerkes. Yerkes Papers. Yale University Library.

———. 1930b, March 14. Letter to Robert Yerkes. Yerkes Papers. Yale University Library.

———. 1930c, March 20. Telegram to Robert Yerkes. Yerkes Papers. Yale University Library.

———. 1930d, April 14. Letter to Robert Yerkes. Yerkes Papers. Yale University Library.

———. 1932. *Gorillas in a native habitat.* Publication No. 426. Washington, DC: Carnegie Institution of Washington.

———. 1933, October 6. Letter to Robert Yerkes. Yerkes Papers. Yale University Library.

———. 1955a, May 24. Letter to Robert Yerkes. Yerkes Papers. Yale University Library.

———. 1955b, July 7. Letter to Robert Yerkes. Yerkes Papers. Yale University Library.

Bingham, Harold Clyde. 1969. *National Cyclopedia of American Biography,* Vol. 51, 367–68. New York: J. T. White.

Birch, H. G. 1945a. The role of motivational factors in insightful problem solving. *Journal of Comparative Psychology* 36:295–314.

———. 1945b. The relation of previous experience to insightful problem-solving. *Journal of Comparative Psychology* 38:367–83.

Birch, H. G., and G. Clark. 1946. Hormonal modification of social behavior. II. The effects of sex-hormone administration on the social dominance status of the female-castrate chimpanzee. *Psychosomatic Medicine* 8:320–31.

———. 1950. Hormonal modification of social behavior. IV. The mechanism of estrogen-induced dominance in chimpanzees. *Journal of Comparative and Physiological Psychology* 43:181–93.

Blakey, A. F. 1976. *Parade of memories: A history of Clay county, Florida.* Jacksonville, FL: Clay County Board of Commissioners.

Blum, D. 1994. *The monkey wars.* New York: Oxford University Press.

Blum, R. A. 1952. Effects of subtotal lesions of frontal granular cortex on delayed reaction in monkeys. *Archives of Neurology and Psychiatry* 67:375–86.

Boring, E., G. 1956. Robert Mearns Yerkes (1876–1956). In Year book of the American Philosophical Society (133–40). Philadeliphia, Pa.: American Philosophical Society.

Bourne, G. H. 1959, October 27. Letter to Leonard Carmichael. Carmichael Papers. American Philosophical Society Archives, Philadelphia.

——— 1971a. *The ape people.* New York: Putnam.

———. 1974. *Primate odyssey.* New York: Putnam.

Boyer, P. 1985. *By the bomb's early light: American thought and culture at the dawn of the atomic age.* New York: Pantheon.

Bruce, D. 1986. Lashley's shift from bacteriology to neuropsychology, 1910–1917, and the influence of Jennings, Watson, and Franz. *Journal of the History of the Behavioral Sciences* 22:27–44.

———. 1991. Integrations of Lashley. In *Portraits of pioneers in psychology* vol. 1, ed. G. A. Kimble, M. Wertheimer, and C. White, 307–23. Washington, DC: APA Books.

———. 1994. Lashley and the problem of serial order. *American Psychologist* 49:93–103.

Bush, V. 1942, April 18. Letter to Dean Francis G. Blake. Files of the Yerkes Regional Primate Research Center. Woodruff Library, Emory University, Atlanta, GA.

———. 1945. *Science—the endless frontier.* Washington, DC: National Science Foundation Reprinted, 1990.

Candland, D. K. 1993. *Feral children and clever animals: Reflections on human nature.* New York: Oxford University Press.

Carmichael, L. 1957, May 6. Letter to Henry Nissen. Carmichael Papers. American Philosophical Society Archives, Philadelphia.

———. 1958, October 24. Letter to Arthur P. Richardson. Carmichael Papers. American Philosophical Society Archives, Philadelphia.

———. 1965. Henry Wieghorst Nissen February 5, 1901–April 27, 1958. *Biographical Memoirs of the National Academy of Sciences of the United States of America* 38:206–22.

———. 1968, December. Some notes on the past, present and future of scientific primatology: Presidential address, Second International Congress of Primatology. Yerkes Regional Primate Research Center, 1–10.

Carpenter, C. R. 1934. A field study of the behavior and social relations of howling monkeys (*Alouatta palliata*). *Comparative Psychology Monographs* 10 (48): 1–168.

———. 1940. A field study in Siam of the behavior and social relations of the gibbon (*Hylobates lar*). *Comparative Psychology Monographs* 16 (84): 1–205.

Cason, H. 1932, March 7. Letter to Robert M. Yerkes. Yerkes Papers. Yale University Library.

Cattell, J. McK. 1902. The Carnegie Institution. *Science* 16:460–69.

———. 1922. The organization of scientific men. *Scientific Monthly* 14:568–77.

Charter of the Yale Laboratories of Primate Biology, Incorporated. 1935. Yerkes Papers. Yale University Library.

Chow, K. L. 1950. A retrograde cell degeneration study of the cortical projection field of the pulvinar in the monkey. *Journal of Comparative Neurology* 93:313–40.

———. 1951a. Numerical estimates of the auditory central nervous system of the rhesus monkey. *Journal of Comparative Neurology* 95:159–75.

———. 1951b. Effects of partial extirpations of the posterior association cortex on visually mediated behavior in monkeys. *Comparative Psychology Monographs* 20 (3): 187–217.

———. 1952. Conditions influencing the recovery of visual discriminative habits in monkeys following temporal neocortical ablations. *Journal of Comparative and Physiological Psychology* 45:430–37.

Chow, K. L., and J. Orbach. 1957. Performance of visual discriminations presented tachistoscopically in monkeys with temporal neocortical ablations. *Journal of Comparative and Physiological Psychology* 50:636–40.

Clark, G., and H. G. Birch. 1945. Hormonal modifications of social behavior. I. The effect of sex-hormones administration on the social status of a male-castrate chimpanzee. *Psychosomatic Medicine* 7:321–29.

———. 1948. Observations on the sex skin and sex cycle in the chimpanzee. *Endocrinology* 43:218–31.

Clark, G., and J. W. Ward. 1949. Responses elicited by combined stimulation of pairs of fixed electrodes in the unanesthetized monkey. *American Journal of Physiology* 158:474–77.

Clarke, A. E. 1987. Research materials and reproductive science in the United States, 1910–1940. In *Physiology in the American context,* ed. G. L. Geison, 323–50. Bethesda, MD: American Physiological Society.

———. 1998. *Disciplining reproduction: Modernity, American life sciences, and "the problems of sex."* Berkeley, CA: University of California Press.

Committee on Animal Models in Biomedical Research. 1995. Aping science: A critical analysis of research at the Yerkes Regional Primate Research Center. *Perspectives on Medical Research* 5:1–102.

Conklin, E. G., F. R. Lillie, and R. M. Yerkes. 1935, August. Report of Committee on Appraisal and Recommendation. Appended to 1936 Annual Report. Yerkes Regional Primate Research Center, Atlanta, GA.

Coolidge, H. J. 1977. Memories of my friend Robert M. Yerkes and his "Prince Chim." In *Progress in ape research,* ed. G. H. Bourne, 43–47. New York: Academic Press.

Corner, G. W. 1937, January 28. Letter to Warren Weaver. Rockefeller Archive Center, North Tarrytown, NY.

———. 1958, August 16. Telegram to Leonard Carmichael. Carmichael Papers. American Philosophical Society Archives, Philadelphia.

———. 1959. Letter to Leonard Carmichael. Corner Papers. American Philosophical Society Archives, Philadelphia.

Cowgill, C. R. 1927, February 22. Comments on report by Dr. H. C. Bingham. Woodruff Library, Emory University, Atlanta, GA.

Cowles, J. T. 1937. Food-tokens as incentives for learning by chimpanzees. *Comparative Psychology Monographs,* 14 (5): 1–96.

Crawford, M. P. 1937. The cooperative solving of problems by young chimpanzees. *Comparative Psychology Monographs* 14 (2): 1–88.

———. 1938. A behavior rating scale for young chimpanzees. *Journal of Comparative Psychology* 26:79–91.

———. 1939. The social psychology of the vertebrates. *Psychological Bulletin* 36:407–46.

———. 1940. The relation between social dominance and the menstrual cycle in female chimpanzees. *Journal of Comparative Psychology* 30:483–513.

———. 1941. The cooperative solving by chimpanzees of problems requiring serial responses to color cues. *Journal of Social Psychology* 13:259–80.

———. 1977. Five years in New Haven and Orange Park: An appreciation. In *Progress in ape research,* ed. G. H. Bourne, 15–20. New York: Academic Press.

Crawford, M. P., J. F. Fulton, C. F. Jacobsen, and J. B. Wolfe. 1948. Frontal lobe ablation in chimpanzee: A resume of "Becky" and "Lucy." In *The Frontal Lobes,* Research Publications of the Association for Research in Nervous and Mental Disease, Vol. 27. Baltimore: Williams & Wilkins.

Crawford, M. P., and K. W. Spence. 1939. Observational learning of discrimination problems by chimpanzees. *Journal of Comparative Psychology* 27:133–47.

Crinella, F. M. 1990. In memoriam: Robert Thompson (1927–1989). *Psychological Science* 1:327–28.

Dagg, A. I. 1984. Sexual bias in the literature of social behaviour in mammals and birds. *International Journal of Women's Studies* 7:118–35.

Danziger, K. 1990. *Constructing the subject: Historical origins of psychological research.* New York: Cambridge University Press.

Davenport, R. K., Jr. 1967. The orangutan in Sabeh. *Folia Primatologica* 5:247–63.

Davenport, R. K., Jr., and G. Berkson. 1963. Stereotyped movements of mental defectives. II. Effects of novel objects. *American Journal of Mental Deficiency* 67:879–82.

Davenport, R. K., Jr., and E. W. Menzel, Jr. 1960. Oddity preference in the chimpanzee. *Psychological Reports* 7:523–26.

———. 1963. Stereotyped behavior of the infant chimpanzee. *Archives of General Psychiatry* 8:99–104.

Davenport, R. K., Jr., E. W., Menzel, Jr., and C. M. Rogers. 1961. Maternal care during infancy: Its effect on weight gain and mortality in the chimpanzee. *American Journal of Orthopsychiatry* 3:803–9.

———. 1966. Effects of severe isolation on "normal" juvenile chimpanzees. *Archives of General Psychiatry* 14:134–38.

Davenport, R. K., Jr., and C. M. Rogers. 1968. Intellectual performance of differentially reared chimpanzees. I. Delayed response. *American Journal of Mental Deficiency* 72:674–80.

———. 1970. Differential rearing of the chimpanzee—a project survey. In *The Chimpanzee,* Vol. 3, *Immunology, Infections, Hormones, Anatomy, and Behavior,* ed. G. H. Bourne, 337–60. Basel: Karger.

DeVore, I., ed. 1965. *Primate behavior: Field studies of monkeys and apes.* New York: Holt, Rinehart and Winston.

de Waal, F. 1982. *Chimpanzee politics: Power and sex among apes.* New York: Harper & Row.

Dewsbury, D. A. 1990. Early interactions between animal psychologists and animal activists and the founding of the APA Committee on Precautions in Animal Experimentation. *American Psychologist* 45:315–27.

———. 1992. Triumph and tribulation in the history of American comparative psychology. *Journal of Comparative Psychology* 106:3–19.

———. 1993. The boys of summer at the end of summer: The Watson-Lashley correspondence of the 1950s. *Psychological Reports* 72:263–69.

———. 1994a. A supplement to Karl Lashley's biographical sketch of Paul Harkai Schiller. *Psychological Record* 44:320–25.

———. 1994b. A history of the Yerkes Laboratories of Primate Biology. I. The Cathy Hayes/Nissen- Robert Yerkes correspondence. Paper presented at meetings of the American Society of Primatologists, Seattle, WA, July, 1994. Abstract in *American Journal of Primatology,* 33:205–6.

———. 1996. Robert M. Yerkes: Psychobiologist with a plan. In *Portraits of pioneers in psychology* Vol. 2, ed. G. A. Kimble, A. Boneau, and M. Wertheimer, 87–105. Washington, DC: APA Books.

———. 1998a. Henry W. Nissen: Quiet comparative psychologist. In *Portraits of pioneers in psychology,* Vol. 3, ed. G. A. Kimble, C. A. Boneau, and M. Wertheimer, 228–42. Washington, DC: APA Books.

———. 1998b. Robert Yerkes, sex research, and the problem of data simplification. *History of Psychology* 1:116–29.

———. 2000. Comparative cognition in the 1930s. *Psychonomic Bulletin & Review* 7:267–83.

———. 2002a. Constructing representations of Karl Spencer Lashley. *Journal of the History of the Behavioral Sciences* 38:225–45.

———. 2002b. The Chicago five: A family group of integrative psychobiologists. *History of Psychology* 5:16–37.

———. 2003a. James Rowland Angell: Born administrator. In *Portraits of pioneers in psychology* Vol. 5, ed. G. A. Kimble and M. Wertheimer, 57–71. Washington, DC: APA Books.

———. 2003b. Conflicting approaches: Operant psychology arrives at a primate laboratory. *The Behavior Analyst* 26:253–65.

———. 2005. Comparative psychology: A case study of development of support for basic research by a federal agency with an applied mission 1948–1963. In *Psychology and the National Institute of Mental Health,* ed. W. E. Pickren and S. F. Schneider. Washington, DC: APA Books.

Diehl, L. A. 1986. The paradox of G. Stanley Hall: Foe of coeducation and educator of women. *American Psychologist* 41:868–78.

Dinsmoor, J. A. 1982. Charles B. Ferster (1922–1981). *American Psychologist* 37:235.

Dodge, R., J. L. Fulton, C. L. Hull, W. R. Miles, and R. P. Angier. 1933, May 29. Letter to Mark A. May. John L. Fulton Papers. Yale University Library.

Drane, H. 1962, Yerkes Lab is most unusual community. *Florida Times-Union,* March 19, 17.

Draper, W. A. 1966. Free-ranging rhesus monkeys: Age and sex differences in individual activity patterns. *Science* 151:476–78.

Dukelow, W. R. 1995. *The alpha males: An early history of the Regional Primate Research Centers.* Lanham, MD: University Press of America.

Elder, J. H. 1934. Auditory acuity of the chimpanzee. *Journal of Comparative Psychology* 17:157–83.

———. 1937a. Methods of anesthetizing chimpanzees. *Journal of Pharmacology and Experimental Therapeutics* 60:347–57.

———. 1937b, October 9. Letter to Robert M. Yerkes. Yerkes Papers. Yale University Library.

———. 1938. The time of ovulation in chimpanzees. *Yale Journal of Biology and Medicine* 10:347–64.

———. 1939a, April. Untitled manuscript. Original in the possession of Ms. Jane Elder Wulff, Battleground, WA.

———. 1939b. Puerperal hemorrhage in chimpanzees. *American Journal of Obstetrics and Gynecology* 37:327–29.

———. 1977. Robert M. Yerkes and memories of early days in the laboratories. In *Progress in ape research,* ed. G. H. Bourne, 29–38. New York: Academic Press.

Elder, J. H., C. G. Hartman, and C. H. Heuser. 1939. A ten and one-half day chimpanzee embryo, *Journal of the American Medical Association* 111:1156–59.

Elliott, R. M. 1956. Robert Mearns Yerkes: 1876–1956. *American Journal of Psychology* 69:487–94.

———. 1958, November 12. Letter to Claire Schiller Lashley. Lashley Papers. Smathers Library, University of Florida.

Embree, E. R. 1925. Memorandum to Mr. Vincent. Rockefeller Archive Center, North Tarrytown, NY.

———. 1926a, April 27. Letter to Robert M. Yerkes. Rockefeller Archive Center, North Tarrytown, NY.

———. 1926b, November 5. Letter to Robert M. Yerkes. Rockefeller Archive Center, North Tarrytown, NY.

———. 1926c, April 30. Letter to Simon Flexner. Rockefeller Archive Center, North Tarrytown, NY.

———. 1929, November 6. Letter to Max Mason. Rockefeller Archive Center, North Tarrytown, NY.

Evarts, E. V. 1952a. Effect of auditory cortex ablation on auditory-visual association in monkey. *Journal of Neurophysiology* 15:435–41.

———. 1952b. Effect of auditory cortex ablation on frequency discrimination in monkey. *Journal of Neurophysiology* 15:443–48.

Falk, J. L. 1958. The grooming behavior of the chimpanzee as a reinforcer. *Journal of the Experimental Analysis of Behavior* 1:83–85.

Fantz, R. L. 1965. Ontogeny of perception. In *Behavior of nonhuman primates: Modern research trends,* Vol. 2, ed. A. M. Schrier, H. F. Harlow, and F. Stollnitz, 365–403. New York: Academic Press.

Ferster, C. B. 1957a. Concurrent schedules of reinforcement in the chimpanzee. *Science* 125:1090–91.

———. 1957b. Withdrawal of positive reinforcement as punishment. *Science* 126:509.

———. 1958a. Intermittent reinforcement of a complex response in a chimpanzee. *Journal of the Experimental Analysis of Behavior* 1:163–65.

———. 1958b. Control of behavior in chimpanzees and pigeons by time out from positive reinforcement. *Psychological Monographs: General and Applied* 72 (8): 1–38.

———. 1959. A complex concurrent schedule of reinforcement. *Journal of the Experimental Analysis of Behavior* 2:65–80.

Ferster, C. B., and B. F. Skinner. 1957. *Schedules of reinforcement.* New York: Appleton Century Crofts.

Finch, G. 1941a. Chimpanzee handedness. *Science* 94:117–18.

———. 1941b. The solution of patterned string problems by chimpanzees. *Journal of Comparative Psychology* 32:83–90.

Finger, S. 1994. *Origins of neuroscience: A history of explorations into brain function.* New York: Oxford University Press.

Fisher, D. 1993. *Fundamental development of the social sciences.* Ann Arbor: University of Michigan Press.

Fletcher, F. 1940. Effects of quantitative variations of food incentive in the performance of physical work by chimpanzees. *Comparative Psychology Monographs* 15 (3):1–46.

Florida Times-Union, 1934. Twin chimpanzees soon to celebrate first birthday. June 13.

Florida Times-Union. 1962a. Bennett says state should buy Yerkes. July 29, 21.

Florida Times-Union. 1962b. Bid by JU to run lab is rejected. March 23, 21.

Florida Times-Union. 1962c. Discuss possible move of Yerkes facility. March 28, 1.

Florida Times-Union. 1962d. Emory studies moving Yerkes lab to Atlanta. March 4, 19.

Foley, B. 1988. Monkey Farm's fame fanned out with hairy arms. *Florida Times-Union,* June 5, F–1, F–4.

Fosdick, R. B. 1952. *The story of the Rockefeller Foundation.* New York: Harper & Brothers.

Frank, L. K. 1928, November 13. Letter to James R. Angell. Rockefeller Archive Center, North Tarrytown, NY.

———. 1935, February 9–16. Diary entry. Rockefeller Archive Center, North Tarrytown, NY.

Fulton, J. F. 1933a, April 5. Letter to Robert M. Yerkes. Yerkes Papers. Yale University Library.

———. 1933b, April 17. Letter to Robert M. Yerkes. Yerkes Papers. Yale University Library.

———. 1933c, June 13. Letter to Robert M. Yerkes. Yerkes Papers. Yale University Library.

———. 1938, February 7. Letter to Stanhope Bayne-Jones. Rockefeller Archive Center, North Tarrytown, NY.

———. 1958, May 1. Letter to C. H. N. Long. Long Papers. American Philosophical Society Archives, Philadelphia.

Furumoto, L. 1989. The new history of psychology. In *The G. Stanley Hall Lecture Series,* Vol. 9, ed. I. S. Cohen, 9–34. Washington, DC: APA Books.

Gardner, L. P., and H. W. Nissen. 1948. Simple discrimination behavior of young chimpanzees: Comparisons with human aments and domestic animals. *Journal of Genetic Psychology* 72:145–64.

Gardner, R. A., and B. T. Gardner, 1969. Teaching sign language to a chimpanzee. *Science* 165:664–72.

Gavan, J. A. 1952. Birth order and birth weight in the chimpanzee. *American Journal of Physical Anthropology* 10:23–30.

———. 1953. Growth and development of the chimpanzee; a longitudinal and comparative study. *Human Biology* 25:93–143.

Gibson, E. J. 1980. Eleanor J. Gibson. In *A history of psychology in autobiography,* Vol. 7, ed. G. Lindzey, 239–71. San Francisco: Freeman.

Gilbert, M. B. F. 1987. Memories of JEAB's mother. *Journal of the Experimental Analysis of Behavior* 48:475–77.

———. 2002a, October 22. Recollections of Yerkes. Unpublished manuscript.

———. 2002b, October 22. Georgia on my mind (excerpt). Unpublished manuscript.

Glickman, S. E. 1996. Donald Olding Hebb: Returning the nervous system of psychology. In *Portraits of pioneers in psychology,* Vol. 2, ed. G. A. Kimble, C. A. Boneau, and M. Wertheimer. Washington, DC: APA Books.

Golarz de Bourne, N., and F. A. King. 1990. Dr. Geoffrey H. Bourne. *Anatomical Record* 228:38A–39A.

Goodall, J. 1986. *The chimpanzees of Gombe: Patterns of behavior.* Cambridge, MA: Belknap.

Goodman, M., and A. J. Riopelle. 1963. Serum transferrins in chimpanzees; inheritance of serum transferrins in chimpanzees. *Nature* 197:261.

Goodwin, W. J., and J. Augustine. 1975. The Primate Research Centers Program of the National Institutes of Health. *Federation Proceedings* 34:1641–42.

Goy, R. W. 1967. William Caldwell Young September 8, 1899–August 30, 1965. *Anatomical Record* 157:3–11.

Gray, G. W. 1955. The Yerkes Laboratories. *Scientific American* 192 (2): 67–77.

Greene, J. D. 1915a, June 7. Letter to Robert M. Yerkes. Rockefeller Archive Center, North Tarrytown, NY.

———. 1915b, November 26. Letter to Robert M. Yerkes. Yerkes Papers. Yale University Library.

Gregg, A. 1935, October 8. Diary entry. Rockefeller Archive Center, North Tarrytown, NY.

———. 1948, April 30. Memorandum concerning an interview with K. S. Lashley. Rockefeller Archive Center, North Tarrytown, NY.

Grether, W. F. 1939. Color vision and color blindness in monkeys. *Comparative Psychology Monographs* 15 (4): 1–38.

Grether, W. F., and R. M. Yerkes. 1940. Weight norms and relations for chimpanzee. *American Journal of Physical Anthropology* 27:181–97.

Griffin, D. R. 1984. *Animal thinking*. Cambridge, MA: Harvard University Press.

Griswold, A. W. 1952, February 28. Letter to James B. Conant. Carmichael Papers. American Philosophical Society Archives, Philadelphia.

Grun, B. 1991. *The timetables of history.* 3rd ed. New York: Simon & Schuster.

Guilloud, N. B. 1993a, November 1. Letter to W. Richard Dukelow.

Guilloud, N. B., and F. L. Fitz-Gerald. 1967. Spontaneous hyperthermia in the gorilla. *Folia Primatologica* 6:177–79.

Guilloud, N. B., and I. C. Kline. 1966. Paralytic poliomyelitis in the gorilla and orangutan. *Physical Therapy* 46:516–18.

Haggerty, M. E. 1909. Imitation in monkeys. *Journal of Comparative Neurology and Psychology* 19:337–455.

Hahn, E. 1988. *Eve and the apes.* New York: Weidenfeld & Nicolson.

Halberstam, D. 1993. *The fifties.* New York: Villard.

Hamilton, G. V. 1911. A study of trial and error reactions in mammals. *Journal of Animal Behavior* 1:33–66.

Hanson, F. B. 1936, October 29–30. Diary entry. Rockefeller Archive Center, North Tarrytown, NY.

———. 1939, December 11. Diary entry. Rockefeller Archive Center, North Tarrytown, NY.

Haraway, D. 1978. Animal sociology and a natural economy of the body politic, Part I: A political physiology of dominance. *Signs: Journal of Women in Culture and Society* 4:21–36.

———. 1983. Signs of dominance: From a physiology to a cybernetics of primate society, C. R. Carpenter, 1930–1970. *Studies in History of Biology* 6:129–219.

———. 1989. *Primate visions: Gender, race, and nature in the world of modern science.* New York: Routledge.

———. 2000. Morphing in the order: Flexible strategies, feminist science studies, and primate revisions. In *Primate encounters: Models of science, gender, and society,* ed. S. C. Strum and L. M. Fedigan, 398–420. Chicago: University of Chicago Press.

Harlow, H. F. 1949. The formation of learning sets. *Psychological Review* 56:51–65.

———. 1958. The nature of love. *American Psychologist* 13:673–85.

Harrell, G. T., Jr. 1964. A research animal farm for a college of medicine. *Journal of Medical Education* 39:459–69.

———. 1980. The first years of the University of Florida College of Medicine. *Journal of the Florida Medical Association* 67:247–250.

Haslerud, G. M. 1977. Reminiscences of the Yerkes Orange Park laboratory in the mid-1930s. In *Progress in ape research,* ed. G. H. Bourne, 21–27. New York: Academic Press.

———. 1979. Introduction to reprint edition of R. M. Yerkes' *The mental life of monkeys and apes,* v–xiv. Delmar, NY: Scholars' Facsimiles and Reprints.

Hawley, E. W. 1992. *The great war and the search for modern order.* 2nd ed. New York: St. Martin's Press.

Hayes, C. 1951. *The ape in our house.* New York: Harper.

Hayes, K. J., and C. Hayes. 1951. The intellectual development of a home-raised chimpanzee. *Proceedings of the American Philosophical Society* 95:105–9.

———, 1954. The cultural capacity of chimpanzee. *Human Biology* 26:288–303.

Hayes, K. J., and C. H. Nissen. 1971. Higher mental functions of a home-raised chimpanzee. In *Behavior of nonhuman primates: Modern research trends,* Vol. 4, ed. A. M. Schrier and F. Stollnitz, 59–115. New York: Academic Press.

Hayes, K. J., R. Thompson, and C. Hayes. 1953. Discrimination learning set in chimpanzees. *Journal of Comparative and Physiological Psychology* 46:99–104.

Hebb, D. O. 1946a. Emotion in man and animal: an analysis of the intuitive processes of recognition. *Psychological Review* 53:88–106.

———. 1946b. On the nature of fear. *Psychological Review* 53:259–76.

———. 1947. Spontaneous neurosis in chimpanzees: Theoretical relations with clinical and experimental phenomena. *Psychosomatic Medicine* 10:3–19.

———. 1949a. Treatment in chimpanzees. I. Method of analysis. *Journal of Comparative Psychology* 42:192–206.

———. 1949b. *Organization of behavior: A neuropsychological theory.* New York: Wiley.

———. 1958, June 23. Letter to Leonard Carmichael. Carmichael Papers. American Philosophical Society Archives, Philadelphia.

———. 1959. Karl Spencer Lashley: 1890–1958. *American Journal of Psychology* 72:142–49.

———. 1980. D. O. Hebb. In *A history of psychology in autobiography,* Vol. 7, ed. G. Lindzey, 273–303. San Francisco: Freeman.

Herschberger, R. 1944, March 2. Letter to Robert M. Yerkes. Yerkes Papers. Yale University Library.

———. 1948. *Adam's rib.* New York: Pellegrini & Cudahy.

Hilgard, E. R. 1965. Robert Mearns Yerkes: May 26, 1876–February 3, 1956. *Biographical Memoirs, National Academy of Sciences* 38:385–425.

Hisaw, F. L. 1962a, April 6. Letter to Leonard Carmichael. Carmichael Papers. American Philosophical Society Archives, Philadelphia.

———. 1962b, October 8. Letter to Leonard Carmichael. Carmichael Papers. American Philosophical Society Archives, Philadelphia.

———. 1962c, December 17. Letter to Leonard Carmichael. Carmichael Papers. American Philosophical Society Archives, Philadelphia.

Hull. C. L. 1932, February 25. Letter to Robert M. Yerkes. Yerkes Papers. Yale University Library.

Jackson, T. A. 1942. Use of the stick as a tool by young chimpanzees. *Journal of Comparative Psychology* 34:223–35.

Jacobsen, C. F. 1932, February 29. Letter to O. L. Tinklepaugh. Papers of the Yerkes Regional Primate Research Center. Woodruff Library, Emory University, Atlanta, GA.

———. 1938a, February 5. Letter to Stanhope Bayne-Jones. Rockefeller Archive Center, North Tarrytown, NY.

———. 1938b, February 14. Letter to Robert M. Yerkes. Rockefeller Archive Center, North Tarrytown, NY.

Jacobsen, C. F., M. M. Jacobsen, and J. G. Yoshioka. 1932. Development of an infant chimpanzee during her first year. *Comparative Psychology Monographs* 9 (1): 1–94.

Jacobsen, C. F., and H. W. Nissen. 1932, February 26. Letter to O. L. Tinklepaugh. Papers of the Yerkes Regional Primate Research Center. Woodruff Library, Emory University, Atlanta.

———. 1937. Studies of cerebral function in primates. IV. The effects of frontal lobe lesions on the delayed alternation habit in monkeys. *Journal of Comparative Psychology* 23:101–12.

Jacobsen, C. F., J. B. Wolfe, and T. A. Jackson. 1935. An experimental analysis of the functions of the frontal association areas in primates. *Journal of Nervous and Mental Disease* 82:1–14.

Kay, L. E. 1993. *The molecular vision of life*. New York: Oxford University Press.

Keefer, L. B. 1958, November 3. Letter to Claire Schiller Lashley. Lashley Papers. Smathers Library, University of Florida.

Kelleher, R. T. 1957. Conditioned reinforcement in chimpanzees. *Journal of Comparative and Physiological Psychology* 50:571–75.

———. 1958. Concept formation in chimpanzees. *Science* 128:777–78.

Kelleher, R. T., and Morse, W. H. 1987. The Yerkes connection. *Journal of the Experimental Analysis of Behavior* 48:456–57.

Keller, F. S. 1981. Charles Bohris Ferster (1922–1981). *Journal of the Experimental Analysis of Behavior* 36:299–301.

Kellogg, W. N., and L. A. Kellogg. 1933. *The ape and the child: A study of environmental influence upon early behavior.* New York: McGraw-Hill.

Kevles, D. J. 1968. Testing the army's intelligence: Psychologists and the military in World War I. *Journal of American History* 55:565–81.

———. 1977. *The physicists: The history of a scientific community in modern America.* Cambridge, MA: Harvard University Press.

———. 1985. *In the name of eugenics*. Berkeley: University of California Press.

———. 1992. Foundations, universities, and trends in support for the physical and biological sciences, 1900–1992. *Daedalus* 121 (4): 195–235.

Kinnaman, A. J. 1902. Mental life of two *Macacus rhesus* monkeys in captivity, I and II. *American Journal of Psychology* 13:98–172, 209–18.

Klüver, H. 1937, February 9. Letter to Warren Weaver. Rockefeller Archive Center, North Tarrytown, NY.

———. 1942a, June 15. Letter to Karl S. Lashley. Files of the Yerkes Regional Primate Research Center. Woodruff Library, Emory University, Atlanta, GA.

Koehler, E. G. 1960, April 18. Letter to Leonard Carmichael. Carmichael Papers. American Philosophical Society, Philadelphia.

Kohler, R. E. 1991. *Partners in Science: Foundations and natural scientists 1900–1945.* Chicago: University of Chicago Press.

Köhler, W. 1925. *The mentality of apes*. London: Routledge & Kegan Paul.

Kohts, N. 1928. Adaptive motor habits of the *Macaca rhesus* under experimental conditions: A contribution to the problem of "labour processes" of monkeys. *Scientific Memoirs of the Museum Darwinianum in Moscow.* In Russian with English summary. Oxford, England: Museum Darwinianum.

Krantz, D. L. 1972. Schools and systems: The mutual isolation of operant and non-operant psychology as a case study. *Journal of the History of the Behavioral Sciences* 8:86–102.

Ladygina-Kohts, N. N. 1935/2002. *Infant chimpanzee and human child: A classic 1935 comparative study of ape emotions and intelligence,* ed. F. B. M. de Waal. Oxford: Oxford University Press.

Lapin, B., and E. Fridman. 1966. *Monkeys for science*. Moscow: Novosti.

Lashley, C. S. 1958, December 12. Letter to Mike Elliot. Lashley Papers. Smathers Library, University of Florida.

Lashley, K. S. n.d. Autobiographical fragment. Archives of the History of American Psychology, University of Akron, Akron, OH.

———. 1923. The behavioristic interpretation of consciousness I and II. *Psychological Review* 30:237–72 and 329–53.

———. 1929. *Brain mechanisms and intelligence*. Chicago: University of Chicago Press.

———. 1930. Basic neural mechanisms in behavior. *Psychological Review* 37:1–24.

———. 1931. Cerebral control versus reflexology: A reply to Professor Hunter. *Journal of General Psychology* 5:3–20.

———. 1937, January 8. Letter to Frank B. Hanson. Rockefeller Archive Center, North Tarrytown, NY.

———. 1938. Experimental analysis of instinctive behavior. *Psychological Review* 45:445–71.

———. 1941a, December 26. Letter to Dean Francis G. Blake. Files of the Yerkes Regional Primate Research Center. Woodruff Library, Emory University, Atlanta, GA.

———. 1941b, December 20. Letter to President James B. Conant Files of the Yerkes Regional Primate Research Center. Woodruff Library, Emory University, Atlanta, GA.

———. 1942a, January 8. Memorandum to Dean Francis K. Blake. Yerkes Papers. Yale University Library.

———. 1942b, January 15. Preliminary statement. Yerkes Papers. Yale University Library.

———. 1945a, June 13. Letter to Dr. T. G. Andrews. Files of the Yerkes Regional Primate Research Center. Woodruff Library, Emory University, Atlanta, GA.

———. 1949. Persistent problems in the evolution of mind. *Quarterly Review of Biology* 24:28–42.

———. 1950a. In search of the engram. In *Physiological mechanisms in animal behaviour*, ed. J. F. Danielli and R. Brown, 454–482. New York: Academic Press.

———. 1950b. Functional interpretation of anatomic patterns. *Proceedings of the Association for Research on Nervous and Mental Diseases* 30:529–47.

———. 1950c. Memorandum to George W. Gray. Files of the Yerkes Regional Primate Research Center. Woodruff Library, Emory University, Atlanta, GA.

———. 1951. The problem of serial order in behavior. In *Cerebral mechanisms in behavior*, ed. L. A. Jeffress, 112–36. New York: Wiley.

———. 1954a, June 16. Letter to John B. Watson. Lashley Papers. Smathers Library, University of Florida.

———. 1954b, February 16. Letter to Karl Pribram. Provided to the author by Dr. Karl Pribram.

———. 1954c, May 23. Letter to Karl H. Pribram. Provided by Dr. Pribram.

———. 1954d, February 13. Letter to Karl Pribram. Provided by Dr. Pribram.

———. 1957a, May 15. Letter to Leonard Carmichael. Carmichael Papers. American Philosophical Society Archives, Philadelphia.

———. 1957b, September 29. Letter to Leonard Carmichael. Carmichael Papers. American Philosophical Society Archives, Philadelphia.

———. 1958a, May 1. Letter to Ada Watterson Yerkes. Yerkes Papers. Yale University Library.

———. 1958b, May 5. Letter to Arthur P. Richardson. Carmichael Papers. American Philosophical Society Archives, Philadelphia.

———. 1958c, May 19. Effects of the growth of Orange Park on the future of the laboratories. Carmichael Papers. American Philosophical Society Archives, Philadelphia.

———. 1958d. Cerebral organization and behavior. Proceedings of the Association for Research on Nervous and Mental Diseases 36:1–18.

———. 1958e, June 7. Letter to Niko Tinbergen. Lashley Papers. Smathers Library, University of Florida.

———. 1994. Paul Harkai Schiller: 1908–1949. *Psychological Record* 44:309–19.

———. 1998. Vanuxem Lectures (1952). In *The neuropsychological theories of Lashley and Hebb: Contemporary perspectives fifty years after Hebb's* The Organization of Behavior, ed. J. Orbach, 275–340. Lanham, MD: University Press of America.

Lashley, K. S., K. L., Chow, and J. Semmes. 1951. An examination of the electrical field theory of cerebral integration. *Psychological Review,* 58:123–36.

Lashley, K. S., and G. Clark. 1946. The cytoarchitecture of the cerebral cortex of *Ateles:* A critical examination of architectonic studies. *Journal of Comparative Psychology* 85:223–306.

Lashley, K. S., and M. Wade. 1946. The Pavlovian theory of generalization. *Psychological Review* 53:72–87.

Laties, V. G. 1986. Lessons from the history of behavioral pharmacology. In *Advances in behavioral pharmacology,* Vol. 5, *Developmental behavioral pharmacology,* ed. N. A. Krasnegor, D. B. Gray, and T. Thompson, 21–39. Hillsdale, NJ: Erlbaum.

———. 1987. Society for the Experimental Analysis of Behavior: The first thirty years (1957–1987). *Journal of the Experimental Analysis of Behavior* 48:495–512.

Lester, C. T. 1958a, April 17. Letter to Leonard Carmichael. Carmichael Papers. American Philosophical Society Archives, Philadelphia.

———. 1958b, May 22. Letter to Donald O. Hebb. Smathers Library, University of Florida, Gainesville, FL.

———. 1958c, June 26. Letter to W. D. Neff. Smathers Library, University of Florida, Gainesville, FL.

———. 1958d, May 30. Letter to Niko Tinbergen. Smathers Library, University of Florida, Gainesville, FL.

———. 1958e, June 9. Letter to Niko Tinbergen. Smathers Library, University of Florida, Gainesville, FL.

———. 1959. Letter to H. Malcolm Gillette. Rockefeller Archive Center, North Tarrytown, NY.

Malinowski, B. 1927. *Sex and repression in savage society.* New York: Harcourt Brace.

Many Friends and Colleagues Provide Graham Townsend Memorial. 1972. *Yerkes Newsletter* 9 (1): 26–27.

Maslow, A. H. 1932, February 11. Letter to Robert M. Yerkes. Yerkes Papers. Yale University Library.

Mason, M. 1928a, October 3. Letter to James R. Angell. Rockefeller Archive Center, North Tarrytown, NY.

———. 1928b, October 12. Diary entry. Rockefeller Archive Center, North Tarrytown, NY.

———. 1929, November 8. Letter to Edwin R. Embree. Rockefeller Archive Center, North Tarrytown, NY.

Mason, W. A. 1964. Sociability and social organization in monkeys and apes. In *Advances in Experimental Psychology,* Vol. 1, ed. L. Berkowitz, 277–305. New York: Academic Press.

———. 1965. Determinants of social behavior in young chimpanzees. In *Behavior of nonhuman primates* vol. 2, ed. A. M. Schrier, H. F. Harlow, and F. Stollnitz, 335–64. New York: Academic Press.

Mason, W. A., and G. Berkson. 1962. Conditions influencing vocal responsiveness of infant chimpanzees. *Science* 137:127–28.

Mason, W. A., R. K. Davenport, Jr., and E. W. Menzel, Jr. 1968. Early experience and the social development of rhesus monkeys and chimpanzees. In *Early experience and behavior,* ed. G. Newton, and S. Levine, 1–41. Springfield, IL: C. C. Thomas.

May, H. F. 1959. *The end of American innocence.* New York: Columbia University Press.

Mayer, A. G. 1914, April 5. Letter to Robert M. Yerkes. Yerkes Papers. Yale University Library.

McBride, A. F., and D. O. Hebb. 1946. Behavior of the captive bottle-nose dolphin, *Tursiops truncatus. Journal of Comparative and Physiological Psychology* 41:111–23.

McCulloch, T. L. 1939. The role of clasping activity in adaptive behavior of the infant chimpanzee: I, II, and III. *Journal of Psychology* 7: 283–316.

———. 1941. Discrimination of lifted weights by chimpanzees. *Journal of Comparative Psychology* 32:507–19.

McGraw, M. B. 1941. A thought on the process of evolution. McGraw Papers. Millbank Memorial Library, Teachers College, Columbia University, New York.

Mendres, K. A., and F. B. M. de Waal. 2000. Capuchins do cooperate: The advantage of an intuitive task. *Animal Behaviour* 60:523–29.

Menzel, E. W., Jr. 1960. Selection of food by size in the chimpanzee and comparison with human judgment. *Science* 131:1527–28.

———. 1961. Perception of food size in the chimpanzee. *Journal of Comparative and Physiological Psychology* 54:588–91.

———. 1964. Responsiveness to object-movement in young chimpanzees. *Behaviour* 24:147–60.

———. 1966. Responsiveness to objects in free-ranging Japanese monkeys. *Behaviour* 26: 130–50.

Menzel, E. W., Jr., R. K. Davenport, Jr., and C. M. Rogers. 1963a. The effects of environmental restriction upon the chimpanzee's responsiveness to objects. *Journal of Comparative and Physiological Psychology* 56:78–85.

———. 1963b. Effects of environmental restriction upon the chimpanzee's responsiveness in novel situations. *Journal of Comparative and Physiological Psychology* 56:329–34.

Menzel, E. W., Jr., and W. A. Draper. 1965a. Effects of size and shape of aperture upon visual exploration by rhesus monkeys. *Perceptual and Motor Skills* 20:581–89.

———. 1965b. Primate selection of food by size: Visible versus invisible rewards. *Journal of Comparative and Physiological Psychology* 59:231–39.

Merriam, J. C. 1927, February 21. Letter to Robert M. Yerkes. Yerkes Papers. Yale University Library.

Miles, W. R. 1938a, January 27. Minutes of the Committee on Yale Laboratories of Primate Biology. Rockefeller Archive Center, North Tarrytown, NY.

———. 1938b, February 24. Minutes of the Committee on Yale Laboratories of Primate Biology. Rockefeller Archive Center, North Tarrytown, NY.

Miller, G. S., Jr. 1928. Some elements of sexual behavior in primates and their possible influence on the beginnings of human social development. *Journal of Mammalogy* 9:273–93.

———. 1931. The primate basis of human sexual behavior. *Quarterly Review of Biology* 6:379–410.

Miller, H. M., Jr. 1936, February 20. Diary entry. Rockefeller Archive Center, North Tarrytown, NY.

Minutes of meeting group of advisers concerning proposal for establishment station for anthropoid research. Woodbridge Hall, Yale University. 1929, February 17. Yerkes Regional Primate Research Center, Atlanta, GA.

Mitchell, R. W. 1999. Scientific and popular conceptions of the psychology of great apes from the 1790s to the 1970s: Déjà vu all over again. *Primate Reports* 53:3–118.

Morawski, J. G. 1986. Organizing knowledge and behavior at Yale's Institute of Human Relations. *Isis* 77:219–42.

———. 1988. Impossible experiments and practical constructions: The social bases of psychologists' work. In *The rise of experimentation in American psychology,* ed. J. Morawski, 72–93. New Haven, CT: Yale University Press.

Morrison, R. S. 1955, May 2. Memorandum of lunch with Mr. Burdock. Rockefeller Archive Center, North Tarrytown, NY.

Mullen, S. A. 1962, May 4. The proposed move of the Yerkes Laboratories of Primate Biology. Carmichael Papers. American Philosophical Society Archives, Philadelphia.

Münsterberg, H. 1902. The Carnegie Institution. *Science* 16:521–24.

Nadler, R. D. 1977. Sexual behavior of the chimpanzee in relation to the gorilla and orangutan. In *Progress in ape research,* ed. G. H. Bourne, 191–206. New York: Academic Press.

———. 1981. Laboratory research on sexual behavior of the great apes. In *Reproductive biology of the great apes: Comparative and biomedical perspectives,* ed. C. E. Graham, 191–248. New York: Academic Press.

Nadler, R. D., J. F. Dahl, D. C. Collins, and K. G. Gould. 1994. Sexual behavior of chimpanzees (*Pan troglodytes*): Male versus female regulation. *Journal of Comparative Psychology* 108:58–67.

Neff, W. D. 1958a, July 30. Letter to Charles T. Lester. Carmichael Papers. American Philosophical Society Archives, Philadelphia.

———. 1958b, August 12. Letter to Charles T. Lester. Carmichael Papers. American Philosophical Society Archives, Philadelphia.

New York Times. 1929. Rockefellers unite two philanthropies. January 4, 27.

New York Times. 1934. Twin chimpanzees now one year old. July 27.

New York Times. 1988. Dr. Geoffrey Bourne, anatomist: Primate expert a prolific writer. July 21.

Newman, B. M. 1939. Dated chimpanzees. *Scientific American* 160 (2): 71–73.

Nielsen, W. A. 1972. *The big foundations.* New York: Columbia University Press.

Nissen, C. 1955, August 10. Letter to the Yerkes family. Provided by the Yerkes Regional Primate Research Center, Atlanta, GA. Woodruff Library, Emory University, Atlanta, GA.

Nissen, H. W. 1931. A field study of the chimpanzee. *Comparative Psychology Monographs* 8 (1): 1–122.

———. 1938, October 3. Letter to Robert M. Yerkes. Yerkes Papers. Yale University Library.

———. 1942. Studies of infant chimpanzees. *Science* 95:159–61.

———. 1951. Phylogenetic comparison. In *Handbook of experimental psychology,* ed. S. S. Stevens, 347–86. New York: Wiley.

———. 1954a. The nature of drive as innate determinant of behavioral organization. *Nebraska Symposium on Motivation* 2:281–321.

———. 1954b, March 19. Letter to Karl Pribram. Provided by Dr. Pribram.

———. 1956, August 12. Letter to Ada W. and Roberta Yerkes. Yerkes Papers. Yale University Library.

———. 1957a, June 12. Letter to Ada W. Yerkes. Yerkes Papers. Yale University Library.

———. 1957b, May 22. Letter to Leonard Carmichael. Carmichael Papers. American Philosophical Society Archives, Philadelphia.

Nissen, H. W., J. S. Blum, and R. A. Blum. 1948. Analysis of matching behavior in chimpanzee. *Journal of Comparative and Physiological Psychology* 48:62–74.

Nissen, H. W., C. R. Carpenter, and J. T. Cowles. 1936. Stimulus- versus response-differentiation in delayed reactions of chimpanzees. *Journal of Genetic Psychology* 48:112–36.

Nissen, H. W., K. L. Chow, and J. Semmes. 1951. Effects of restricted opportunity for tactual, kinesthetic, and manipulative experience on the behavior of a chimpanzee. *American Journal of Psychology* 64:485–507.

Nissen, H. W., and M. P. Crawford. 1936. A preliminary study of food-sharing behavior in young chimpanzees. *Journal of Comparative Psychology* 22:383–419.

Nissen, H. W., and C. F. Jacobsen. 1932a, May 16. Letter to Robert M. Yerkes. Yerkes Papers. Yale University Library.

———. 1932b, June 25. Letter to Robert M. Yerkes. Yerkes Papers. Yale University Library.

Nissen, H. W., A. H. Riesen, and V. Nowlis. 1938. Delayed response and discrimination learning by chimpanzees. *Journal of Comparative Psychology* 26:361–86.

Nowlis, V. 1941. The relation of degree of hunger to competitive interactions in chimpanzee. *Journal of Comparative Psychology* 32:91–115.

———. 1942. Sexual status and degree of hunger in chimpanzee competitive interaction. *Journal of Comparative Psychology:* 34:185–94.

———. 1994a, October. Memories of life with the Yale chimpanzees. Unpublished manuscript.

O'Donnell, J. M. 1985. *The origins of behaviorism: American psychology, 1870–1920.* New York: New York University Press.

Orange Park, Florida, and the move to Atlanta. Winter, 20–23. *Inside Yerkes.*

Orbach, J. 1959. "Functions" of striate cortex and the problem of mass action. *Psychological Bulletin* 56:271–92.

———. 1982. *Neuropsychology after Lashley: Fifty years since the publication of Brain Mechanisms and Intelligence.* Hillsdale, NJ: Erlbaum.

———. 1998. *The neuropsychological theories of Lashley and Hebb: Contemporary perspectives fifty years after* Hebb's The Organization of Behavior. Lanham, MD: University Press of America.

Orbach, J., and G. J. Fischer. 1959. Bilateral resections of frontal granular cortex. *Archives of Neurology* 1:78–86.

Osterbrock, D. E. 1997. *Yerkes Observatory, 1892–1950: The birth, near death, and resurrection of a scientific research institution.* Chicago: University of Chicago Press.

Pearce, R. M. 1928, May 7. Letter to Robert M. Yerkes. Rockefeller Archive Center, North Tarrytown, NY.

Phillips, H. M. 1955a, October 26. Letter to Robert M. Yerkes. Yerkes Papers. Yale University Library.

———. 1955b, July. Report on the Yerkes Laboratories of Primate Biology, Incorporated. Yerkes Papers. Yale University Library.

Pickren, W. E. 1997. Robert Yerkes, Calvin Stone, and the beginning of programmatic sex research by psychologists, 1921–1930. *American Journal of Psychology* 110:603–19.

Pool, J. L. 1954, May 20. Letter to Leonard Carmichael. Carmichael Papers. American Philosophical Society Archives, Philadelphia.

Pribram, K. H. 1954a, March 10. Letter to H. W. Nissen. Provided by Dr. Pribram.

———. 1954b, May 25. Letter to Karl Lashley. Provided by Dr. Pribram.

———. 1955, February 9. Letter to Karl Lashley. Provided Dr. Pribram.

Pribram, K. H., K. L. Chow, and J. Semmes. 1953. Limit organization of the cortical projection from the medial thalamic nucleus in monkey. *Journal of Comparative Neurology* 98:443–48.

Puente, A. E. 1995. Roger Wolcott Sperry (1913–1994). *American Psychologist* 50:940–41.

Rappleye, W. C. 1955. At the quarter-century mark. In *The Josiah Macy, Jr. Foundation: A review of activities.* New York: Macy Foundation.

Reitz, J. W. 1962, December 21. Letter to J. B. Culpepper. Reitz Papers. University of Florida Libraries, Gainesville, FL.

Rhea, M. 1976. Orange Park real estate developer monkeys with Yerkes laboratory. *Florida Times-Union,* November 21, B7.

Richardson, A. P. 1959a, October 30. Letter to the Board of Scientific Directors. Carmichael Papers. American Philosophical Society Archives, Philadelphia.

———. 1959b, December 10. Application for Research Grant. Carmichael Papers. American Philosophical Society Archives, Philadelphia.

Riesen, A. H. 1950. Arrested vision. *Scientific American* 183 (1): 16–19.

———. 1971. Nissen's observations on the development of sexual behavior in captive-born, nursery-reared chimpanzees. In *The chimpanzee,* Vol. 4, ed. G. Bourne, 1–18. Basel: Karger.

———. 1977. Introduction. In *Progress in ape research,* ed. G. H. Bourne, 1–4. New York: Academic Press.

Riesen, A. H., and E. F. Kinder. 1952. *Postural development of infant chimpanzees.* New Haven, CT: Yale University Press.

Riopelle, A. J. 1959a, October 17. Letter to J. Franklin Yeager. Carmichael Papers. American Philosophical Society Archives, Philadelphia.

———. 1959b, October 27. Letter to J. Franklin Yeager. Carmichael Papers. American Philosophical Society Archives, Philadelphia.

———. 1959c, July 14. Letter to Ada W. Yerkes. Yerkes Papers. Yale University Library.

———. 1960a, January 13. Letter to Leonard Carmichael. Carmichael Papers. American Philosophical Society Archives, Philadelphia.

———. 1960b, February 5. Letter to Leonard Carmichael. Carmichael Papers. American Philosophical Society Archives, Philadelphia.

———. 1961, June 23. Letter to Leonard Carmichael. Carmichael Papers. American Philosophical Society Archives, Philadelphia.

———. 1962a, June 19. Letter to Leonard Carmichael. Carmichael Papers. American Philosophical Society Archives, Philadelphia.

———. 1962b, December 20. Letter to J. Wayne Reitz. Reitz Papers. University of Florida Libraries, Gainesville, FL.

———. 1962c. Some behavioral effects of ionizing radiation on primates. In *Response of the Nervous System to Ionizing Radiation,* ed. R. S. Snider and T. J. Haley, 719–28. New York: Academic Press.

———. 1963a. Growth and behavioral changes in chimpanzees. *Zeitschrift für Morphologie und Anthropologie* 53:53–61.

———. 1963b. Chimpanzee-associated hepatitis. In *Proceedings of a Conference on Research with Primates,* ed. D. E. Pickering, 19–26. Beaverton, OR: Tektronix Foundation.

Riopelle, A. J., U. Rahm, N. Itoigawa, and W. A. Draper. 1964. Discrimination of mirror-image patterns by rhesus monkeys. *Perceptual and Motor Skills* 19:383–89.

Riopelle, A. J., and C. M. Rogers. 1965. Age changes in chimpanzees. In *Behavior of nonhuman primates: Modern research trends,* Vol. 2, ed. A. M. Schrier, H. F. Harlow, and F. Stollnitz, 449–62. New York: Wiley.

Robinson, F. G. 1992. *Love's story told: A life of Henry Murray.* Cambridge, MA: Harvard University Press.

Roe, A. 1952a. *The making of a scientist.* New York: Dodd, Mead.

———. 1952b. Lashley Folder. Anne Roe Papers. Archives of the American Philosophical Society, Philadelphia.

Rogers, C. M. 1957, June 4. Letter to Frank A. Beach. Provided by C. M. Rogers.

Rogers, C. M., and R. K. Davenport, Jr. 1969a. Effects of restricted rearing on sexual behavior of chimpanzees. *Developmental Psychology* 1:200–204.

———. 1969b. Sexual behavior of differentially-reared chimpanzees. In *Proceedings of the Second International Congress of Primatology, Atlanta, GA 1968,* Vol. 1, *Behavior,* ed. C. R. Carpenter, 173–77. Basel: Karger.

———. 1970. Chimpanzee maternal behavior. In *The chimpanzee,* Vol. 3, ed. G. H. Bourne, 361–68. Basel: Karger.

Rumbaugh, D. M. 1998. Austin H. Riesen (1913–1996). *American Psychologist* 53:60–61.

Samelson, F. 1979. Putting psychology on the map: Ideology and intelligence testing. In *Psychology in social context,* ed. A. R. Buss, 103–68. New York: Irvington.

Samuel S. Fels Fund. 1963. A report for the two-year period ended December 31, 1963. Philadelphia: Samuel S. Fels Fund.

Schiller, P. H. 1949. Delayed detour response in the octopus. *Journal of Comparative and Physiological Psychology* 42:220–25.

———. 1951. Figural preferences in the drawings of a chimpanzee. *Journal of Comparative and Physiological Psychology* 44:101–11.

———. 1952. Innate constituents of complex responses in primates. *Psychological Review* 59:177–91.

———. 1957. Innate motor action as a basis of learning: Manipulative patterns in the chimpanzee. In *Instinctive behavior: The development of a modern concept,* ed. C. H. Schiller, 264–87. New York: International Universities Press.

Schmidt, G. H. W. 1962, April 6. Memorandum. Reitz Papers. University of Florida Libraries, Gainesville, FL.

———. 1963, March 28. Letter to Justine C. Ward. Woodruff Library, Emory University, Atlanta, GA.

Schusterman, R. J. 1962. Transfer effects of successive discrimination-reversal training in chimpanzees. *Science* 137:422–23.

———. 1964a. The effect of a manipulatory incentive on social interactions of rhesus monkeys. *Animal Behaviour* 12:416–19.

———. 1964b. Strategies of normal and mentally retarded children under conditions of uncertain outcome. *American Journal of Mental Deficiency* 69:66–75.

———. 1964c. Successive discrimination-reversal training and multiple discrimination training in one-trial learning by chimpanzees. *Journal of Comparative and Physiological Psychology* 58:153–56.

Schusterman, R. J., and I. S. Bernstein. 1962. Response tendencies of gibbons in single and double alternation tasks. *Psychological Reports* 11:521–22.

Scrapbook. n.d. Scrapbook of press clippings maintained at the Yerkes Regional Primate Research Center and the Woodruff Library, Atlanta, GA.

Scripps, E. W. 1915, November 4. Letter to Robert M. Yerkes. Yerkes Papers. Yale University Library.

Seymour, C. 1938a, April 13. Letter to Warren Weaver. Rockefeller Archive Center, North Tarrytown, NY.

———. 1938b, October 7. Letter to Warren Weaver. Rockefeller Archive Center, North Tarrytown, NY.

———. 1939, January 4. Letter to Warren Weaver. Rockefeller Archive Center, North Tarrytown, NY.

Sidman, M. 1960. *Tactics of scientific research: Evaluating experimental data in psychology.* New York: Basic Books.

Skinner, B. F. 1981. Charles B. Ferster—A personal memoir. *Journal of the Experimental Analysis of Behavior* 35:259–60.

Spence, K. W. 1934. Visual acuity and its relation to brightness in chimpanzee and man. *Journal of Comparative Psychology* 18:333–61.

———. 1936. The nature of discrimination learning in animals. *Psychological Review* 43:427–49.

———. 1939. The solution of multiple choice problems by chimpanzees. *Comparative Psychology Monographs* 15 (3): 1–54.

Spence, K. W., and R. M. Yerkes. 1937. Weight, growth and age in chimpanzee. *American Journal of Physical Anthropology,* 22:229–46.

Sperry, R. W. 1943. Effect of 180 degree rotation of the retinal field on visuo-motor coordination. *Journal Experimental Zoology* 92:263–79.

———. 1945a. The problem of central nervous reorganization after nerve regeneration and muscle transposition. *Quarterly Review of Biology* 20:311–69.

———. 1945b. Centripetal regeneration of the 8th cranial nerve root with systematic restoration of vestibular reflexes. *American Journal of Physiology* 144:735–41.

———. 1947. Nature of functional recovery following regeneration of the oculomotor nerve in amphibians. *Anatomical Record* 97:293–316.

Spragg, S. D. S. 1940. Morphine addiction in chimpanzees. *Comparative Psychology Monographs* 15 (7): 1–132.

Temerlin, M. K. 1975. *Lucy: Growing up human.* Palo Alto, CA: Science and Behavior Books.

Teuber, M. L. 1994. The founding of the Primate Station, Tenerife, Canary Islands. *American Journal of Psychology* 107:551–81.

Thompson, N. S. 1925, May 28. Letter to James R. Angell. Rockefeller Archive Center, North Tarrytown, NY.

———. 1929a, January 25. Letter to Edmund E. Day. Rockefeller Archive Center, North Tarrytown, NY.

———. 1929b, May 23. Letter to James R. Angell. Rockefeller Archive Center, North Tarrytown, NY.

———. 1939, January 25. Letter to Charles Seymour. Rockefeller Archive Center, North Tarrytown, NY.

———. 1942, April 8. Letter to President Seymour. Files of the Yerkes Regional Primate Research Center. Woodruff Library, Emory University, Atlanta, GA.

Thorndike, E. L. 1911. *Animal intelligence.* New York: Macmillan.

Thorne, B. M. 1995. Robert Thompson: Karl Lashley's heir? *Journal of the History of the Behavioral Sciences* 31:129–36.

Thorpe, W. H. 1979. *The origins and rise of ethology: The science of the natural behaviour of animals.* London: Praeger.

Time. Babe and ape. June 19, 1933. 44.

Tinbergen, N. 1958, June 4. Letter to Dr. C. T. Lester. Lashley Papers. University of Florida.

Tinklepaugh, O. L. 1931. A diet for chimpanzees and monkeys in captivity. *Journal of Mammalogy* 12:400–404.

———. 1932a. Multiple delayed reaction with chimpanzees and monkeys. *Journal of Comparative Psychology* 13:207–43.

———. 1932b, December 10. Letter to Robert M. Yerkes. Yerkes Papers. Yale University Library.

———. 1932c, March 2. Letter to C. F. Jacobsen and H. W. Nissen. Papers of the Yerkes Regional Primate Research Center. Woodruff Library, Emory University, Atlanta, GA.

———. 1932d, July 7. Letter to H. W. Nissen and C. Jacobsen. Papers of the Yerkes Regional Primate Research Center. Woodruff Library, Emory University, Atlanta, GA.

———. 1933a. Corrections to "A diet for chimpanzees and monkeys in captivity." *Journal of Mammalogy* 14:68–69.

———. 1933b. Sex cycles and other cyclic phenomena in a chimpanzee during adolescence, maturity, and pregnancy. *Journal of Morphology* 54:521–46.

Titchener, E. B. 1902. The Carnegie Institution. *Science* 16:609–11.

Tomasello, M, and J. Call. 1997. *Primate cognition*. New York: Oxford University Press.

Tomilin, M. I., and R. M. Yerkes. 1935. Chimpanzee twins: Behavioral relations and development. *Journal of Genetic Psychology* 46:237–63.

Triplet, R. G. 1992. Henry A. Murray: The making of a psychologist? *American Psychologist* 47:299–307.

Valenstein, E. S. 1973. *Brain control: A critical examination of brain simulation and psychosurgery*. New York: Wiley.

Vincent, G. E. 1926, November 17. Letter to Robert M. Yerkes. Rockefeller Archive Center, North Tarrytown, NY.

———. 1929, May 31. Letter to the Yale anthropoids. Rockefeller Archive Center, North Tarrytown, NY.

———. 1952. Behavioral effects of prefrontal lobectomy, lobotomy and circumsection in the monkey (*Macaca mulatta*). *Journal of Comparative Neurology* 96:179–207.

Warren, J. M., and K. Akert. 1960. Impaired problem solving by cats with thalamic lesions. *Journal of Comparative and Physiological Psychology* 53:207–8.

Watson, J. B. 1906. The need of an experimental station for the study of certain problems in animal behavior. *Psychological Bulletin* 3:149–56.

———. 1907, May 29. Letter to Robert M. Yerkes. Yerkes Papers. Yale University Library.

———. 1908. Imitation in monkeys. *Psychological Bulletin* 5:169–78.

———. 1916a, November 29. Letter to Robert M. Yerkes. Cedric Larson Papers. Archives of the History of American Psychology, Akron, OH.

———. 1916b, May 12. Letter to Robert M. Yerkes. Yerkes Papers. Yale University Library.

Weaver, W. 1935, March 30. Diary entry. Group 1.1, series 2000, Box 165, Folder 2024, Rockefeller Archive Center, North Tarrytown, NY.

———. 1936, November 25. Diary entry. Rockefeller Archive Center, North Tarrytown, NY.

———. 1937a, January 3. Diary entry. Rockefeller Archive Center, North Tarrytown, NY.

———. 1937b, January 9. Diary entry. Rockefeller Archive Center, North Tarrytown, NY.

———. 1937c, January 14. Diary entry. Rockefeller Archive Center, North Tarrytown, NY.

———. 1937d, January 25. Diary entry. Rockefeller Archive Center, North Tarrytown, NY.

———. 1937e, February 17. Diary entry. Rockefeller Archive Center, North Tarrytown, NY.

———. 1937f, February 17. Diary entry. Rockefeller Archive Center, North Tarrytown, NY.

———. 1937g, September 14. Diary entry. Rockefeller Archive Center, North Tarrytown, NY.

———. 1938a, February 17. Diary entry. Rockefeller Archive Center, North Tarrytown, NY.

———. 1938b, March 8. Diary entry. Rockefeller Archive Center, North Tarrytown, NY.

———. 1938c, December 8. Letter to Charles Seymour. Rockefeller Archive Center, North Tarrytown, NY.

———. 1939, December 14. Diary entry. Rockefeller Archive Center, North Tarrytown, NY.

———. 1940, February 19. Letter to Charles Seymour. Rockefeller Archive Center, North Tarrytown, NY.

———. 1955, April 12. Diary entry. Rockefeller Archive Center, North Tarrytown, NY.

Weidman, N. 1999. *Constructing scientific psychology: Karl Lashley's mind-brain debates.* Cambridge: Cambridge University Press.

Welker, W. I. 1956a. Some determinants of play and exploration in chimpanzees. *Journal of Comparative and Physiological Psychology* 49:84–89.

———. 1956b. Effects of age and experience on play and exploration of young chimpanzees. *Journal of Comparative and Physiological Psychology* 49:223–26.

Wiener, A. S., J. A. Gavan, and E. B. Gordon. 1953. Blood group factors in anthropoid apes and monkeys. II. Further studies on the Rh-Hr factors. *American Journal of Physical Anthropology:* 11:39–45.

Wiener, A. S., J. A. Gavan, and E. B. Gordon. 1958. Blood group factors in anthropoid apes and monkeys. II. Further studies on the Rh-Hr factors. *American Journal of Physical Anthropology* 11:39–45.

Wiener, A. S., J. Moor-Jankowski, and C. M. Rogers. 1965. Heredity of blood groups of chimpanzees. *American Journal of Human Genetics* 17:495–99.

Wight, R. D., and K. A. Smith. 1998. Making place for a comparative scientist: Robert Mearns Yerkes at Harvard, 1902–1917. *International Journal of Comparative Psychology* 11:13–143.

Williams, G. 1947. What we can learn from the apes. *Saturday Evening Post, August 2,* 26–96.

Winston, A. S. 1996. "As his name indicates": R. S. Woodworth's letters of reference and employment for Jewish psychologists in the 1930s. *Journal of the History of the Behavioral Sciences* 32: 30–43.

Witmer, L. 1909. A monkey with a mind. *Psychological Clinic* 3:179–205.

Wolfe, J. B. 1936. Effectiveness of token-rewards for chimpanzees. *Comparative Psychology Monographs* 12 (5): 1–72.

Wolfle, D. 1946. The reorganized American Psychological Association. *American Psychologist* 1:3–6.

Yale Anthropoids. 1929, May 27. Memorandum to the Rockefeller Philanthropists. Rockefeller Archive Center, North Tarrytown, NY.

Yeager, J. F. 1968. *Regional Primate Research Centers: The creation of a program.* Bethesda, MD: National Institutes of Health.

Yerkes, A. W. 1939. Yale in Florida. Unpublished manuscript. Yerkes Papers. Yale University Library.

———. 1956, February 4. Telegram to Leonard Carmichael. Carmichael Papers. American Philosophical Society Archives, Philadelphia.

———. 1957. Family matters: A record. Yerkes Papers Yale University Library.

Yerkes, R. M. 1899. Reaction of Entomostraca to stimulation by light. *American Journal of Physiology* 3:157–82.

———. 1901. The formation of habits in the turtle. *Popular Science Monthly* 58:519–25.

———. 1907. *The dancing mouse: A study in animal behavior.* New York: Macmillan.

———. 1909. Scientific method in animal psychology. *Extrait des Comptes rendus du VI Congres international de Psychologie, Geneva,* 809–810.

———. 1911. *Introduction to psychology.* New York: Holt Rinehart.

———. 1913a, May 20. Letter to Alfred G. Mayer. Yerkes Papers. Yale University Library.

———. 1914a, February 25. Letter to Abraham Flexner. Yerkes Papers. Yale University Library.

———. 1914b. The study of human behavior. *Science* 39:625–33.

———. 1914c. The Harvard laboratory of animal psychology and the Franklin Field Station. *Journal of Animal Behavior* 4:176–84.

———. 1915a. Letter to Abraham Flexner. Rockefeller Archive Center, North Tarrytown, NY.

———. 1915b. Progress and peace. *Scientific Monthly* 1:195–201.

———. 1915c. Letter to Jerome D. Greene. Rockefeller Archive Center, North Tarrytown, NY.

———. 1916a, January 11. Letter to Simon Flexner. Yerkes Papers. Yale University Library.

———. 1916b, May 15. Letter to the President and Trustees of the Carnegie Corporation. Yerkes Papers. Yale University Library.

———. 1916c. Provision for the study of monkeys and apes. *Science* 43:231–34.

———. 1916d. A new method for studying ideational and allied forms of behavior in man and other animals. *Proceedings of the National Academy of Sciences, USA* 2:631–33.

———. 1916e. The mental life of monkeys and apes: A study of ideational behavior. *Behavior Monographs* 3 (12): 1–145.

———. 1916f, May 16. Letter to John B. Watson. Yerkes Papers. Yale University Library.

———. 1919. Diary entry. Yerkes Papers. Yale University Library.

———. 1921a. Psychological examining in the United States Army. *Memoirs of the National Academy of Sciences* 15:1–890.

———. 1921b. Memorandum on psychological research prepared with special reference to The Johns Hopkins University. Yerkes Papers. Yale University Library.

———. 1922, June 11. Letter to J. R. Angell. Yerkes Papers. Yale University Library.

———. 1923a. Testing the human mind. *Atlantic Monthly* 131:358–70.

———. 1923b, January 29. Letter and proposal to J. R. Angell. Yerkes Papers. Yale University Library.

———. 1924, February 12. Letter and proposal to J. R. Angell. Yerkes Papers. Yale University Library.

———. 1925a. *Almost human.* London: Jonathan Cape.

———. 1925b, January 19. Letter and proposal to John C. Merriam. Rockefeller Archive Center, North Tarrytown, NY.

———. 1925c, April 14. Letter and proposal to Edwin R. Embree. Rockefeller Archive Center, North Tarrytown, NY.

———. 1925d, May 18. Letter to Edwin R. Embree. Rockefeller Archive Center, North Tarrytown, NY.

———. 1926a, February 17. Letter to E. W. Scripps. Yerkes Papers. Yale University Library.

———. 1926b, November 9. Letter to Edwin R. Embree. Rockefeller Archive Center, North Tarrytown, NY.

———. 1926c, May 6. Letter to Edwin R. Embree. Rockefeller Archive Center, North Tarrytown, NY.

———. 1926d. Annual report of activities of Robert M. Yerkes and associated investigators in Institute of Psychology, Yale University, for the academic year 1925–26. Rockefeller Archive Center, North Tarrytown, NY.

———. 1927a. A program of anthropoid research. *American Journal of Psychology* 39:181–99.

———. 1927b. The mind of a gorilla. *Genetic Psychology Monographs* 2 (1, 2): 1–193.

———. 1927c. Third annual report of research activities of Robert M. Yerkes and associated investigators in Institute of Psychology, Yale University, for the academic year 1926–27. Rockefeller Archive Center, North Tarrytown, NY.

———. 1927d. The mind of a gorilla. Part II. Mental development. *Genetic Psychology Monographs* 2 (6): 375–551.

———. 1927e. Summary report of progress of studies in psycho-biological problems of sex in infrahuman organisms at the Institute of Psychology, Yale University, for the year 1926–27. National Academy of Sciences Archives, Washington, DC.

———. 1928a. The mind of a gorilla. Part III. Memory. *Comparative Psychology Monographs* 5 (2): 1–92.

———. 1928b, October 18. Letter and memorandum to Max Mason. Rockefeller Archive Center, North Tarrytown, NY.

———. 1928c, November 7. Letter and memorandum to Lawrence K. Frank. Rockefeller Archive Center, North Tarrytown, NY.

———. 1929a, October 31. Memorandum on the Medical Laboratories of Pasteur Institute for utilization of African primates at Kindia, French Guiana. Yerkes Papers. Yale University Library.

———. 1929b, February 12. Informational memorandum concerning feasibility of plan for establishment of anthropoid breeding and observation station. Rockefeller Archive Center, North Tarrytown, NY.

———. 1929c, April 6. Informational memorandum on location and development of anthropoid breeding and observation station. Rockefeller Archive Center, North Tarrytown, NY.

———. 1929d. Log and notes of Robert M. Yerkes for the African portion of his European-African journey, June–October, 1929. Yerkes Papers. Yale University Library.

———. 1929e, March 15. Letter to James Rowland Angell. Yerkes Papers. Yale University Library.

———. 1929f. Plan for the development of the Department of Comparative Psycho-Biology, the Institute of Human Relations, Yale University. Yerkes Papers. Yale University Library.

———. 1930a. Robert Mearns Yerkes: Psychobiologist. In *A history of psychology in autobiography,* vol. 2, ed. C. Murchison, 381–407. Worcester, MA: Clark University Press.

———. 1930b, March 21. Letter to J. R. Angell. Yerkes Papers. Yale University Library.

———. 1930c, March 13. Letter to R. P. Angier. Yerkes Papers. Yale University Library.

———. 1931a. Foreword to H. W. Nissen's "A Field Study of the Chimpanzee." *Comparative Psychology Monographs* 8 (1): i–vi.

———. 1931b, September 17. Letter to Raymond B. Fosdick. Rockefeller Archive Center, North Tarrytown, NY.

———. 1932a. Yale Laboratories of Comparative Biology. *Comparative Psychology Monographs* 8 (3): 1–33.

———. 1932b. *Methods and Problems of Medical Education.* 20th Series, 1–3. Yale University School of Medicine Laboratories of Comparative Psychobiology.

———. 1932c, December 13. Letter to Otto L. Tinklepaugh. Yerkes Papers. Yale University Library.

———. 1932d, March 19. Letter to Abraham H. Maslow. Yerkes Papers. Yale University Library.

———. 1932e, May 26. Letter to Henry W. Nissen and Carlyle F. Jacobsen. Yerkes Papers. Yale University Library.

———. 1933a, October 11. Letter to Harold Bingham. Yerkes Papers. Yale University Library.

———. 1933b, June 16. Letter to Otto L. Tinklepaugh. Yerkes Papers. Yale University Library.

———. 1933c, March 29. Letter to James R. Angell. Yerkes Papers. Yale University Library.

———. 1933d, April 10. Letter to James R. Angell. Yerkes Papers. Yale University Library.

———. 1933e, May 6. Letter to R. P. Angier. Yerkes Papers. Yale University Library.

———. 1933f, June 2. Letter to James R. Angell. Yerkes Papers. Yale University Library.

———. 1933g, June 6. Letter to James R. Angell. Yerkes Papers. Yale University Library.

———. 1933h, June 9. Letter to James R. Angell. Yerkes Papers. Yale University Library.

———. 1933i, June 16. Letter to James R. Angell. Yerkes Papers. Yale University Library.

———. 1934a, November 10. Letter to Max Mason. Rockefeller Archive Center, North Tarrytown, NY.

———. 1934b. Modes of behavioral adaptation in chimpanzee to multiple-choice problems. *Comparative Psychology Monographs* 10 (1): 1–108.

———. 1936a, August 11. Letter to Bishop John Dallas. Yerkes Papers. Yale University Library.

———. 1936b. A chimpanzee family. *Journal of Genetic Psychology* 48:362–70.

———. 1937a. Regulations of Yale Laboratories of Primate Biology, Inc. New Haven, CT, 1937.

———. 1937b, December 15. Letter to Charles Seymour. Rockefeller Archive Center, North Tarrytown, NY.

———. 1937c, May 7. Letter to James R. Angell. Yerkes Papers. Yale University Library.

———. 1937d, April 28. Letter to Alan Gregg. Archives, National Library of Medicine, Bethesda, MD.

———. 1937e. Primate cooperation and intelligence. *American Journal of Psychology* 50:254–70.

———. 1937f, October 22. Letter to James H. Elder. Yerkes Papers. Yale University Library.

———. 1938, April 11. Letter to Stanhope Bayne-Jones. Yerkes Papers. Yale University Library.

———. 1939a. Sexual behavior in the chimpanzee. *Human Biology* 11:78–111.

———. 1939b. Social dominance and sexual status in the chimpanzee. *Quarterly Review of Biology* 14:115–36.

———. 1940. Social behavior of chimpanzees: Dominance between mates, in relation to sexual status. *Journal of Comparative Psychology* 30:147–86.

———. 1943. *Chimpanzees: A laboratory colony.* New Haven, CT: Yale University Press.

———. 1944, March 30. Letter to Ruth Herschberger. Yerkes Papers. Yale University Library.

———. 1950. *The scientific way* (also known as *Testament*). Unpublished manuscript. Yerkes Papers. Yale University Library.

———. 1955a, November 4. Letter to H. M. Phillips. Yerkes Papers. Yale University Library.

———. 1955b, June 10. Letter to Leonard Carmichael. Carmichael Papers. American Philosophical Society, Philadelphia, PA.

———. 1955c, March 28. Letter to Harold Bingham. Yerkes Papers. Yale University Library.

———. 1955d, June 6. Letter to Harold Bingham. Yerkes Papers. Yale University Library.

———. 1963. Creating a chimpanzee colony. *Yale Journal of Biology and Medicine* 36:205–23.

Yerkes, R. M., and J. H. Elder. 1936a. Oestrus, receptivity, and mating in chimpanzee. *Comparative Psychology Monographs* 13 (5): 1–39.

———. 1936b. The sexual reproductive cycles of chimpanzee. *Proceedings of the National Academy of Sciences USA* 22:276–83.

Yerkes, R. M., and B. W. Learned. 1925. *Chimpanzee intelligence and its vocal expressions.* Baltimore: Williams & Wilkins.

Yerkes, R. M., and S. D. S. Spragg. 1937. La mesure du comportement adapte chez les chimpanzes. *Jurnal de Psycholgie Normale et Pathologique* 34:449–74.

Yerkes, R. M., and M. I. Tomilin. 1935. Mother-infant relations in chimpanzee. *Journal of Comparative Psychology* 20:321–58.

Yerkes, R. M., and J. B. Watson. 1911. Methods of studying vision in animals. *Behavior Monographs* 1 (2): 1–90.

Yerkes, R. M., and A. W. Yerkes. 1929. *The great apes: A study of anthropoid life.* New Haven, CT: Yale University Press.

Young, W. C., and W. D. Orbison. 1944. Changes in selected features of behavior in pairs of oppositely sexed chimpanzees during the sexual cycle and after ovariectomy. *Journal of Comparative Psychology* 37:107–43.

Zuckerman, S. 1932. *The social life of monkeys and apes.* New York: Harcourt Brace.
———. 1978. *From apes to warlords.* New York: Harper & Row.

Personal Communications

Anderson, M. B. 1993, August 6. Letter to Donald A. Dewsbury.

Berkson, G. 1993, October 18. Letter to D. A. Dewsbury.

———. 1996, October 31. Interview with D. A. Dewsbury Chicago, IL.

Bernstein, I. S. 1993, November 5. Interview with D. A. Dewsbury. Athens, GA.

Blanshard, R. Y. 1993, August 6. Interview with D. A. Dewsbury. Hamden, CT.

Carpenter, J. T. 1993, September 21. Letter to D. A. Dewsbury.

Crawford, M. P. 1993, December 27. Interview with Donald A. Dewsbury, Cosmos Club, Washington, DC.

Daniel, R. S. 1994. Letter to D. A. Dewsbury.

Draper, W. A. 1994, September 2. Audiotape made for D. A. Dewsbury.

Falk, J. L. 1994, May 28. Interview with D. A. Dewsbury, Atlanta, GA.

Gilbert, M. B. 2002c, October 31. Electronic mail message to D. A. Dewsbury.

Green, B. 1994, April 26. Interview with D. A. Dewsbury.

Guilloud, N. B. 1993b, November 19. Letter to D. A. Dewsbury.

———. 2002, October 26. electronic mail message to D. A. Dewsbury.

Hebb, J. 2004, July 30. Telephone conversation with D. A. Dewsbury.

Itoigawa, N. 1994, October 27. Letter to D. A. Dewsbury.

Jackson, T. A. 1993, October. Audiotape made for D. A. Dewsbury.

Mason, W. A. 1994, July 27. Interview with D. A. Dewsbury, Seattle, WA.

Menzel, E. W., Jr. 1994, May 26 and 26. Interview with D. A. Dewsbury. East Setauket, NY.

Moor-Jankowski, J. 1994, April 21. Letter to Donald A. Dewsbury.

Nowlis, V. 1994b, April 20. Interview with D. A. Dewsbury, Orange Park, FL.

Peacock, L. J. 1993, November 5. Interview with D. A. Dewsbury. Athens, GA.

Pollard, R. 1994. Interview with Donald A. Dewsbury and Rene Wilkie, Orange Park, FL.

Pribram, K. H. 1993, April 23 Interview with Donald A. Dewsbury, Clearwater, FL.

———. 1994 June 15. Letter to Donald A. Dewsbury.

Riesen, A. H., and and Riesen, H. 1994, August 8. Interview with Donald A. Dewsbury, Riverside, CA.

Riesen, H. 1994, September 3. Notes on recollections of Orange Park. Personal communication to D. A. Dewsbury.

Riopelle, A. J. 1994, January 24. Interview with D. A. Dewsbury, Gainesville, FL.

Roberts, Judy. 1993, November 4. Interview with D. A. Dewsbury, Atlanta.

Rogers, C. M. 1994, February 2. Interview with D. A. Dewsbury. Worthington Springs, FL.

Schusterman, R. J. 1993, September 16. Audiotape made for D. A. Dewsbury.

Spragg, S. D. S. 1993, September. Audiotape made for Donald A. Dewsbury.

Tigges, J., and Tigges, M. 1993, November 4. Interview with D. A. Dewsbury Atlanta.

Warren, H. B. 1994, February 15. Telephone conversation with D. A. Dewsbury.

Welker, W. I. 1996, May 29. Yerkes Laboratories of Primate Biology Orange Park FL: Recollections of Wally Welker from 1952–54. Electronic mail communication to D. A. Dewsbury.

Wilhite, M. 1994, February 23. Interview with D. A. Dewsbury, Orange Park, FL.

Yerkes, D. N. 1993, December 27. Interview with Donald A. Dewsbury, Washington, DC.

Index of Names

Page numbers in italics refer to illustrations or tables.

Subject Index

Page numbers in italics refer to illustrations or tables.